PRACTICAL VHDL

An Introduction

PRACTICAL VHDL

An Introduction

Blaine C. Readler

Full Arc

PRACTICAL VHDL
AN INTRODUCTION

Published by Full Arc Press

Visit us at: http://www.readler.com

E-mail: blaine@readler.com

ISBN: 978-0-9992296-8-2

Printed in the United States of America

First Edition: 2022

Dedicated to the memory of my father, Clifford, the ultimate example of practical humanity.

He who would learn to fly one day must first learn to stand and walk and run and climb and dance; one cannot fly into flying.
—Friedrich Nietzsche

Contents

Introduction

Rather than explain the VHDL language bit-by-bit from the bottom up (or top down), this book assumes that you are adept at applying and learning from examples. The goal of the book is to get you up and running, creating useful designs in the simplest, straightest path possible. What it does not do, is provide you a comprehensive understanding of the vast scope of the language proper. If you want to become an expert at VHDL, this book is a good place to start, but you will eventually need to take on the whole breadth of the subject via one of the many good thick (and more expensive) texts.

The book assumes that you already understand basic digital concepts, like binary and hexadecimal number systems, and the use of registers in implementing designs.

In order to benefit completely from the examples in this book, you might consider having the following:
- a computer (of course);
- a text editor program;
- FPGA compiler software;
- VHDL simulator;

When choosing a text editor, it's important to make sure that the program is "VHDL aware," meaning that the editor understands basic VHDL syntax, and uses color to show the various parts. The examples in this book use the Notepad++ program, available for free here:

https://notepad-plus-plus.org/downloads/

FPGA vendors (e.g., Intel and Xilinx) provide compiling software to use in the development of their FPGA products. Most vendors allow free downloads of their compiling programs that can be used for their smaller FPGA devices. Note that their compiling programs create programming files (the information loaded into the FPGA to make it run) that are specific to their FPGA devices.

Some vendor compiling software includes simulation software as well. For example, the Xilinx Vivado compiling software includes embedded simulation, and when downloading the Intel/Altera Quartus compiler, you are given (as of this writing) the option to also download a starter version of the tried-and-true Modelsim simulation program.

Ultimately, you may want to load an actual FPGA device and watch your design operate. There are a number of vendors offering hobby-class development boards. For example, Digilent offers the Arty-S7 (currently ~$100).

All code examples used in this book are available as text files at:

http://www.readler.com

A note about punctuation: commas and periods are generally placed before closing parenthesis. For example, the following words might describe my approach to writing this book: "fastidious," "thorough," and "clarity." However, I have taken the liberty to break this rule in order to avoid confusion about the exact spelling of signal names. So, for example, in this context I might write that "in_1", "out_1", and "enable_b" comprise all the signals of block "mux_2".

Finally, I use the term "function" sometimes, and you should be aware that VHDL offers a specific type of code element called a "function," which is beyond the scope of this book.

Chapter 1

VHDL Files

We begin by introducing a basic VHDL file.

```vhdl
-- ------------------------------------------------------------
--    Useful information goes here at the top, like Copyright
--    notices, authors, creation dates, etc..  All of this is
--    ingnored by the compiler.
-- ------------------------------------------------------------

library IEEE;
use IEEE.STD_LOGIC_1164.all;
use IEEE.NUMERIC_STD.all;
use IEEE.STD_LOGIC_MISC.all;
use IEEE.STD_LOGIC_UNSIGNED.all;


entity led_blinker is
  port
    (
    clk              : in  std_logic; --50MHz
    --
    led              : out std_logic
    );
end entity led_blinker;

architecture Behavioral of led_blinker is

    signal counter          : unsigned(27 downto 0);

begin

    led_counter : process(clk)
    begin
      if rising_edge(clk) then
        counter <= counter + 1;
      end if;
    end process;

    led <= counter(27);

end architecture Behavioral;
```

Fig. 1-1

As with all FPGA languages, a VHDL compiler expects certain very specific structures. All VHDL design files include four basic sections, as shown here:

```
  ------------------------------------------------------------------
--  1  Useful information goes here at the top, like Copyright
--     notices, authors, creation dates, etc..  All of this is
--     ingnored by the compiler.
  ------------------------------------------------------------------
```

```
library IEEE;
use IEEE.STD_LOGIC_1164.all;       2
use IEEE.NUMERIC_STD.all;
use IEEE.STD_LOGIC_MISC.all;
use IEEE.STD_LOGIC_UNSIGNED.all;
```

```
entity led_blinker is
  port                             3
  (
    clk             : in  std_logic; --50MHz
    --
    led             : out std_logic
  );
end entity led_blinker;
```

```
architecture Behavioral of led_blinker is

    signal counter          : unsigned(27 downto 0);

begin

    led_counter : process(clk)
    begin                          4
      if rising_edge(clk) then
        counter <= counter + 1;
      end if;
    end process;

    led <= counter(27);

end architecture Behavioral;
```

Fig. 1-2

1) An informational header section comes first (actually optional). Double dashes (--) indicate comments, and the compiler ignores everything that follows on that line.

2) Like many software program languages, VHDL assumes that you are using built-in libraries (built in to the complier, that is). These are the most common.

3) Here we begin to see some meat. For our purposes (an introduction to VHDL designs), we can think of an entity as essentially the design that we'll eventually be loading into an FPGA. This is the entity declaration section. The words "entity" "is" and "port" as well as "end entity" are key words, and are always required. Here, "led_blinker" is the name that we've given to this entity (our design). Although not required by VHDL, you will want to name your file the same as the entity. The lines between the parenthesis establish ("declare") the input and output signals of the entity, in our case, the I/O pins of the FPGA. We'll dissect this in more detail soon.

4) Finally, we come to the actual design details, the "architecture." VHDL allows more than one architecture to be associated with an entity, but for our purposes, there will always be one architecture for each design (entity). Again, we'll dive into the details of this soon.

Entity declarations

Here's the details for the entity declaration section:

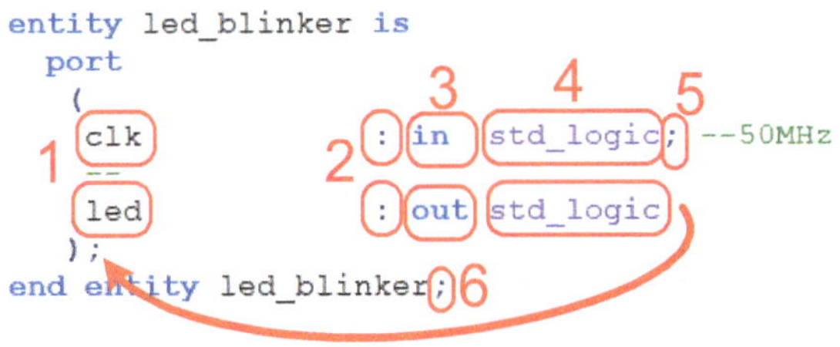

Fig. 1-3

1) "clk" and "led" are the I/O signal names for the entity (the design). VHDL is not case sensitive (doesn't care regarding upper case/lower case usage) in signal names, but you'd be prudent to be consistent. Typically, regular signals are all lower case, while special uses, e.g., constants, might be all-caps. You'll occasionally find designers using a mix of lower and upper case (called camel case).

2) The declared signal names are always followed by a colon,

3) and these are then followed by an indication of the direction of the signal, whether it's an input ("in"), or an output ("out"), or possibly bi-directional ("inout", both an input and output).

4) Finally, the compiler needs to know the type of the signal, where "type" has a formal meaning in VHDL. The "std_logic" indicates that these signals are of the Standard Logic type (the most common by far), and are defined in the IEEE.STD_LOGIC_1164 library shown previously. Although we will encounter other types as we go along, all I/O signals at the entity level will be of the std_logic (and the later-introduced std_logic_vector) type.

5) Each declaration line ends with a semi-colon. Except, that is, for the last one, where that semi-colon mysteriously moves outside the parenthesis.

6) A semi-colon also marks the end of the entire entity declaration.

Architecture

And, here's the architecture components:

```
architecture Behavioral of led_blinker is

declaration {   signal counter          : unsigned(27 downto 0);

        begin

            led_counter : process(clk)
            begin
                if rising_edge(clk) then
    body {          counter <= counter + 1;
                end if;
            end process;

            led <= counter(27);

        end architecture Behavioral;
```

Fig. 1-4

The keyword "architecture" is followed by the name that we give it. It could be anything, but "behavioral" is commonly used, since it delineates between other types in much more sophisticated design scenarios. The "led_blinker" tells the compiler to which entity this architecture belongs (seemingly redundant in our examples, since our architecture and entity sections are always together in the same file). Note that "architecture", "of", and "is" are all necessary key words.

The architecture is divided into two basic sections, the "declaration" section, where we declare (establish) the signals and other value-holding things such as constants that we will be using in the design, and the "body," which includes all the functional details of the design. The key word "begin" marks the beginning of the architecture body, and the architecture as a whole ends with the key words "end architecture" and the name that we've given it. Note the final semi-colon.

Here are the details of the architecture contents:

```
architecture Behavioral of led_blinker is
    signal counter       : unsigned(27 downto 0);
begin
    led_counter : process(clk)
    begin
        if rising_edge(clk) then
            counter <= counter + 1;
        end if;
    end process;
    led <= counter(27);
end architecture Behavioral;
```

Fig. 1-5

1) Each signal declaration begins with the key word "signal", followed by the name of the signal, in this case "counter".

2) As in the entity declaration section, a colon follows the name.

3) Here we declare the formal type of the signal. It could be the "std_logic" type (or "std_logic_vector"), but in this case is "unsigned" since, as we'll see, we use it arithmetically. As it happens, our "counter" signal is 28 bits wide, and, whereas you are probably familiar with something like "[27:0]" to indicate this, in VHDL the method is much more verbose. "(27 downto 0)" must serve our purpose.

4) Again, as in the entity declaration section, the line ends with a semi-colon.

5) These group of lines make up what we call a process, and we'll dive into this as we describe the operation of our little design.

6) This is an example of a direct signal assignment, where we assign the MSB (most significant bit) of the "counter" signal to our "led" output. Once again, note that a semi-colon marks the end of the line.

Blinking LED

This first simple design blinks an LED approximately every five seconds. The operation consists of a continually cycling counter, whose most significant bit drives the LED. The counter is implemented in this process:

```
1  led_counter : process(clk)
   begin
       if rising_edge(clk) then
4          counter <= counter + 1;
       end if;
   end process;
```

Fig. 1-6

A process is a fundamental VHDL design structure, whereby we tell the compiler what inputs to be sensitive to, and then define a variety of logical operations on one or more signals within the process. By sensitive inputs, we mean that any changes on the defined sensitivity signals "wake up" the process. Perhaps the most common, and most useful, type of process is one which implements one or more registers, as in this case. Let's look at the elements and how they create a register(s).

1) This is the name of this process. It can be whatever you like, as long as no other process has the same one.

2) A colon separates the process name from the required key word "process."

3) The key word is followed by the sensitivity signal list, bounded by parenthesis. The compiler ignores the entire process unless something in the sensitivity list changes. In this case, we have just one signal, the clock. So the only time the process is activated is when the clock signal either rises or falls. This, of course, begins to sound like the operation of a clocked register.

4) The key words "begin", and "end process", followed by a semi-colon mark the beginning and end of the process.

Which leaves us with the core, the functional part, of the process:

```
led_counter : process(clk)
begin
   if rising_edge(clk) then
      counter <= counter + 1;
   end if;
end process;
```

Fig. 1-7

1) An if-then conditional statement should be somewhat familiar if you've done any type of programming. Only if the conditions are met are the following lines performed.

2) The key words "end if" mark the end of the conditionally processed lines, followed by the nearly ubiquitous semi-colon.

Here we find the conditional test, something that must be true to continue:

```
led_counter : process(clk)
begin      1            2
   if rising_edge(clk) then
      counter <= counter + 1;
   end if;
end process;
```

Fig. 1-8

1) Along the way, VHDL introduced the "rising_edge" key word, which is essentially a convenient way to tell the compiler that this is indeed a regular clocked register we're implementing. The lines that follow are processed only when a rising edge of …

2) the "clk" signal is seen. Remember that the "clk" signal in the sensitivity list limited entry into the process to changes to "clk", and this key word further limits operation to just the rising edge half. The key word "falling_edge" is also valid, and, of course, implements a negative edge-clocked register.

We finally come to the input and output of this clocked register:

```
led_counter : process(clk)
begin
   if rising edge(clk) then
   1  counter <= counter + 1;
   end if;
end process;
```

Fig. 1-9

1) The output of the register is the signal "counter".

2) The greater-than/equal sign ("<=") is the VHDL assignment indication. Whatever is to the left of this is the target of the assignment (what gets assigned), and whatever is to the right is the source of the assignment (the thing that gets passed on to the target). Because this is a clocked process, what's to the right gets assigned to the "counter" signal at each "clk" signal rising edge.

3) In this case, the source of the assignment is the target signal "counter" itself, but incremented by one (+1). Each "clk"

clock cycle, the value of "counter" increases by one, which, of course, implements a binary counter.

4) Assignment lines always end with a semi-colon.

Remember that the signal type of "counter" was "unsigned". Since we are adding something (one) to "counter", its signal type must be able to handle arithmetic, and "unsigned" is the most convenient arithmetic type to use in this case. (You might also recall that one of the included libraries was "IEEE.STD_LOGIC_UNSIGNED.all", which is necessary when using unsigned type signals).

We saw that the "counter" signal is 28 bits wide, meaning that it is a 28-bit vector signal (versus a single-bit scalar). A 28-bit binary signal has a range from zero to 2^{28}-1, or 0 to 268,435,455. This means that the most significant bit (27) toggles once each complete cycle of the counter, or, since our clock "clk" is indicated (via comments) to be 50MHz, the MSB toggles once each 20 nanoseconds-times-268,435,455, or about once each 5.37 seconds.

This in turn means that, since we are assigning the counter MSB to the LED output signal, then the LED blinks every 5.37 seconds.

```
begin

  led_counter : process(clk)
  begin
    if rising_edge(clk) then
      counter <= counter + 1;
    end if;
  end process;

  led <= counter(27);

end architecture Behavioral;
```

Fig. 1-10

If instead we used bit 26 like so:

led <= counter(26);

then the LED would blink at half the rate, or once each ~2.7 seconds.

And if we used bit 25:
```
led <= counter(25);
```
then the LED would blink at one-quarter the rate, or once each ~1.34 seconds.

Chapter 2

Push-button

The previous example had no FPGA inputs, other than the clock. The LED merrily blinks away undisturbed for as long as power is supplied. Next, we will provide a means to turn the blinking LED on and off, and we'll do this using a push-button switch.

This is the basic idea:

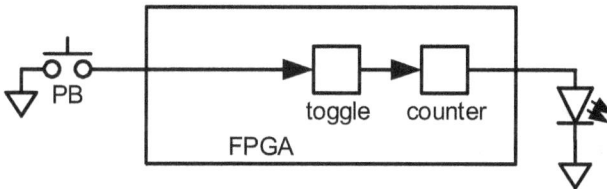

Fig. 2-1

Each time the push-button switch is pressed, the enable to the counter is toggled. Press once, and the LED begins blinking; press again, and the LED turns off.

Any time that we're dealing with mechanical switches, however, we need to take into account the phenomenon of bounce. Although to our human senses, flipping (or pushing) a switch is an instantaneous contact, in fact, the contact points can, and do, bounce a bit before settling into continuous conduction. Although more prevalent on making contact, bounce can occur when lifting the switch as well. Since the contact pieces are so small, the bouncing happens very quickly, over a period of milliseconds. Implemented as just shown, the LED counter would be enabled and disabled a few times before continuing undisturbed. In this instance, this millisecond-long bouncing on/off enabling would not be noticeable. However, since we're toggling off of the switch, we can't know how many times the switch contacts might bounce. If they bounce an even number of times, the toggle will end up where it started, and the effect will be that we never activated the switch.

The next figure shows our solution.

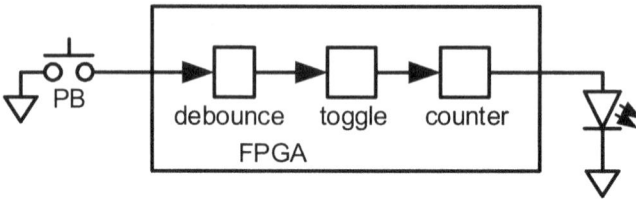

Fig. 2-2

Debounce

We insert a debounce function. This is a sort of logical filter, blocking the various bounces, and letting through just one activation event.

This shows the operation of the debounce.

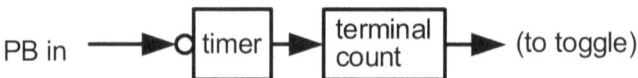

Fig. 2-3

With each bounce of the button, a timer begins anew, and after the last bounce, after the finger is lifted from the button, the final timer countdown reaches the terminal count, making sure the bouncing is done before kicking the toggle function. The period of disabling time must be long enough to mask the longest possible bounce time, but not so long as to get in the way of subsequent legitimate switch activations.

Here's the top part of the implemented VHDL file (the entity declaration):

```
|-- ------------------------------------------------------------
--    Useful information goes here at the top, like Copyright
--    notices, authors, creation dates, etc.. All of this is
--    ingnored by the compiler.
|-- ------------------------------------------------------------

library IEEE;
use IEEE.STD_LOGIC_1164.all;
use IEEE.NUMERIC_STD.all;
use IEEE.STD_LOGIC_MISC.all;
use IEEE.STD_LOGIC_UNSIGNED.all;

entity push_button is
  port
  (
    clk             : in  std_logic; --50MHz
    pb_in           : in  std_logic; --low-active
    --
    led             : out std_logic
  );
end entity push_button;
```

Fig. 2-4

The only difference so far between this and the original LED blinker is that we've added a push-button input, "pb_in". Notice that the comment indicates that it is low-active, since the actual push-button connects the FPGA input pin to ground (logic zero) when pushed. The host circuit board presumably includes a pull-up resistor (so that it's high when not pushed).

Push-button architecture

Here's the architecture:

```vhdl
architecture Behavioral of push_button is

    -- ~1 million
    constant DEBOUNCE      : unsigned(23 downto 0) := X"100000";

    signal pb_count        : unsigned(23 downto 0) := X"000000";
    signal pb_d1           : std_logic := '1';
    signal pb_toggle       : std_logic := '0';
    signal counter_led     : unsigned(27 downto 0) := X"0000000";

begin
    pb_process : process(clk)
    begin
        if rising_edge(clk) then
            pb_d1 <= pb_in;  --synchronize
            --
            if (pb_d1 = '0') then
                pb_count <= DEBOUNCE;
            elsif (pb_count /= X"000000") then
                pb_count <= pb_count - 1;
            end if;
        end if;
    end process;

    pb_toggler : process(clk)
    begin
        if rising_edge(clk) then
            if ( pb_count = X"000001") then
                pb_toggle <= not pb_toggle;
            end if;
        end if;
    end process;

    led_counter : process(clk)
    begin
        if rising_edge(clk) then
            counter_led <= counter_led + 1;
        end if;
    end process;

    led <= pb_toggle AND counter_led(27);

end architecture Behavioral;
```

Fig. 2-5

We've added a good bit of logic, and we'll go through it in functional parts.

This first section includes:

```vhdl
architecture Behavioral of push_button is

    -- ~1 million
    constant DEBOUNCE     : unsigned(23 downto 0) := X"100000";

    signal pb_count       : unsigned(23 downto 0) := X"000000";
    signal pb_d1          : std_logic := '1';
    signal pb_toggle      : std_logic := '0';
    signal counter_led    : unsigned(27 downto 0) := X"0000000";

begin
    pb_process : process(clk)
    begin
        if rising_edge(clk) then
            pb_d1 <= pb_in;   --synchronize
            --
            if (pb_d1 = '0') then
                pb_count <= DEBOUNCE;
            elsif (pb_count /= X"000000") then
                pb_count <= pb_count - 1;
            end if;
        end if;
    end process;

    pb_toggler : process(clk)
    begin
        if rising_edge(clk) then
            if ( pb_count = X"000001") then
                pb_toggle <= not pb_toggle;
            end if;
        end if;
    end process;

    led_counter : process(clk)
    begin
        if rising_edge(clk) then
            counter_led <= counter_led + 1;
        end if;
    end process;

    led <= pb_toggle AND counter_led(27);

end architecture Behavioral;
```

1 { (bracket around led_counter process)

2 (circle around pb_toggle)

Fig. 2-6

1) The original LED blinking counter, unchanged from the first design,

2) except that now we're gating the MSB of the counter with a new signal, "pb_toggle". As we'll soon see, this is the toggling signal that changes with each push of the button to turn on or off the blinking LED.

Blaine C. Readler

PB process

Next, we look at the "pb_process" block.

```vhdl
architecture Behavioral of push_button is

3  -- ~1 million
   constant DEBOUNCE       : unsigned(23 downto 0) := X"100000";

   signal pb_count         : unsigned(23 downto 0)  := X"000000";
   signal pb_d1            : std_logic := '1';
   signal pb_toggle        : std_logic := '0';
   signal counter_led      : unsigned(27 downto 0)  := X"0000000";

begin
   pb_process : process(clk)
   begin
      if rising edge(clk) then
1       pb_d1 <= pb_in;    --synchronize
        --
2       if (pb_d1 = '0') then
           pb_count <= DEBOUNCE;
4       elsif (pb_count /= X"000000") then
           pb_count <= pb_count - 1;
        end if;
      end if;
   end process;

   pb_toggler : process(clk)
   begin
      if rising_edge(clk) then
         if ( pb_count = X"000001") then
            pb_toggle <= not pb_toggle;
         end if;
      end if;
   end process;

   led_counter : process(clk)
   begin
      if rising_edge(clk) then
         counter_led <= counter_led + 1;
      end if;
   end process;

   led <= pb_toggle AND counter_led(27);

end architecture Behavioral;
```

Fig. 2-7

1) Before using the "pb_in" signal internally, we need to synchronize it to our local clock. The reason for this will be clear soon;

2) If the button input is pushed (active low), then we load a counter ("pb_count") with a value named "DEBOUNCE". Note that pb_count is a 24-bit vector signal. If pb_in were not synchronized, then it might conceivably transition just as the local clock was rising. For a single bit (such as pb_d1) this is not that important, since if the register doesn't catch pb_in on this rising edge, then it will catch it on the next. For a multi-bit counter, however, this could cause havoc, since some bits of the counter might transition to the DEBOUNCE value, while other may not, resulting in a scrambled counter load;

3) The value of DEBOUNCE is defined in the signal declaration section via a constant. If you're not familiar with constants, these are values that are (surprise) simply constant, not changed by logical operations. We're using all caps simply as a convenient marker (i.e., all caps will mean that the value is a constant). Note that we've defined DEBOUNCE as "unsigned". This is necessary, since in line 2) we are assigning it to an unsigned counter value, and VHDL requires assignments to be of like types. The ":=" colon/equal sign is the VHDL method of assigning a value to a constant (or declared signal). We are defining DEBOUNCE to be hex 100000. The VHDL language uses this form, X"100000", to define values in hex format (this is equivalent to the 0x100000 form of most computer programming languages). Since hex 100000 is approximately decimal one million, we can see that the counter is loaded with a value that represents about 20 milliseconds (1×10^6 x 20 nanoseconds).

4) In line 2), the pb_counter is continually loaded with DEBOUNCE as long as the button is held pushed, and begins counting down once lifted, freezing when it reaches zero approximately 20 milliseconds later. Note that the counter is reloaded each time the button bounces active. The debounce holdoff time is thus the defined time starting from the last bounce contact after the button is lifted.

Toggler

Now that we have a debounce holdoff (pb_counter), we can let that drive the toggle function.

```vhdl
architecture Behavioral of push_button is

    -- ~1 million
    constant DEBOUNCE      : unsigned(23 downto 0) := X"100000";

    signal pb_count        : unsigned(23 downto 0) := X"000000";
    signal pb_d1           : std_logic := '1';
    signal pb_toggle       : std_logic := '0';
    signal counter_led     : unsigned(27 downto 0) := X"0000000";

begin
    pb_process : process(clk)
    begin
        if rising_edge(clk) then
            pb_d1 <= pb_in;  --synchronize
            --
            if (pb_d1 = '0') then
                pb_count <= DEBOUNCE;
            elsif (pb_count /= X"000000") then
                pb_count <= pb_count - 1;
            end if;
        end if;
    end process;

    pb_toggler : process(clk)
    begin
        if rising_edge(clk) then
            if ( pb_count = X"000001") then
                pb_toggle <= not pb_toggle;
            end if;
        end if;
    end process;

    led_counter : process(clk)
    begin
        if rising_edge(clk) then
            counter_led <= counter_led + 1;
        end if;
    end process;

    led <= pb_toggle AND counter_led(27);

end architecture Behavioral;
```

Fig. 2-8

1) We wait for the holdoff pb_counter to finish its down-count, and let the conditional decision be made on a value of one,

since once the down-count is complete, the counter sits idle at a value of zero. We want to know that the counter *has* finished, not that it *is* continually finished;

   2) And, once the holdoff count has completed, we toggle, i.e., invert, the value of pb_toggle, which we saw turns on and off the blinking LED output.

   A couple of points before we move on:

```vhdl
architecture Behavioral of push_button is

    -- ~1 million
    constant DEBOUNCE      : unsigned(23 downto 0) := X"100000";

    signal pb_count        : unsigned(23 downto 0) := X"000000";
    signal pb_d1           : std_logic := '1';
    signal pb_toggle       : std_logic := '0';
    signal counter_led     : unsigned(27 downto 0) := X"0000000";

begin
```

Fig. 2-9

   1) Just as we are able to assign a value to our constant DEBOUNCE, we can also assign values to signals as well. Unlike constants, though, the value assigned to a signal is not, well, constant. What this means is that the value we assign is simply the initial value of the signal. This value remains until some logic changes it. In the case of pb_count, the initial zero value remains until the first push button activation;

   2) In the case of scalar signals, for example pb_d1, we are assigning an initial single-bit value (here a one). In VHDL single-bit values are denoted with single quotes, as in '1'. Vector values use double quotes, for example "000" is used for a vector signal that was defined as a three-bit value.

   So, for example:

VHDL	decimal
'0'	0
'1'	1
"01"	1
"11"	3
X"01"	1
X"12"	18

Adjusted PB detection

Notice that the LED toggles on/off *after* the user lifts their finger from the push button. Users might find this odd, expecting the LED blinking to go on or off as soon as the button is pushed. We can implement this more usual operation by detecting and toggling when the button is first pushed, and then blocking further toggles until a period of time after the final bounce.

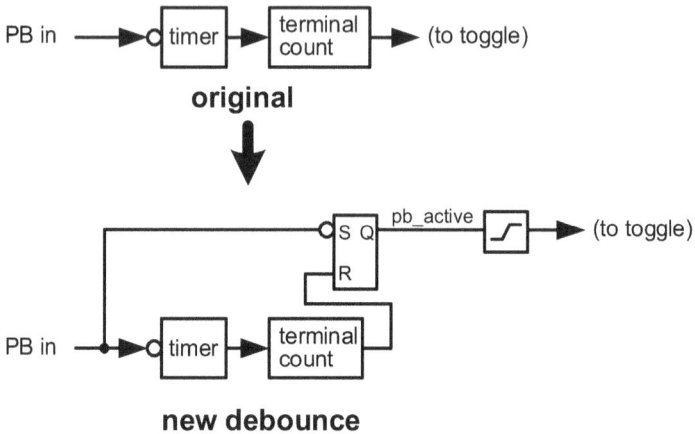

Fig. 2-10

When the push button is pushed, it now not only kicks off the timer, but also sets the set/reset latch "pb_active", which is left activated until the end of the whole pushed-button event is over. We then detect that pb_active latch has been set (that's a "rising edge detect" symbol), and let that toggle the blinking LED on/off state. As in the original implementation, the timer starts over with each bounce, and also while the button is held down, so the debounce filtering extends through and beyond the last lifted bounce.

Here's the new VHDL file, push_button_2:

```vhdl
architecture Behavioral of push_button_2 is

    -- ~1 million
    constant DEBOUNCE     : unsigned(23 downto 0) := X"100000";

    signal counter_led    : unsigned(27 downto 0) := X"0000000";
    signal pb_count       : unsigned(23 downto 0) := X"000000";
    signal pb_d1          : std_logic := '1';
    signal pb_active      : std_logic := '0';
    signal pb_active_d1   : std_logic := '0';
    signal pb_toggle      : std_logic := '0';

begin
    pb_process : process(clk)
    begin
        if rising_edge(clk) then
            pb_d1 <= pb_in;  --synchronize
            --
            if (pb_d1 = '0') then
                pb_count <= DEBOUNCE;
            elsif (pb_count /= X"000000") then
                pb_count <= pb_count - 1;
            end if;
            --
            if (pb_d1 = '0') then
                pb_active <= '1';
            elsif (pb_count = X"000001") then
                pb_active <= '0';
            end if;
            -- rising edge detection and toggle
            pb_active_d1 <= pb_active;
            if ( pb_active = '1' AND pb_active_d1 = '0' ) then
                pb_toggle <= not pb_toggle;
            end if;
        end if;
    end process;

    led_counter : process(clk)
    begin
        if rising_edge(clk) then
            counter_led <= counter_led + 1;
        end if;
    end process;

    led <= pb_toggle AND counter_led(27);

end architecture Behavioral;
```

Fig. 2-11

First, we can identify portions carried over from the previous design.

```vhdl
architecture Behavioral of push_button_2 is

    -- ~1 million
    constant DEBOUNCE      : unsigned(23 downto 0) := X"100000";

    signal counter_led     : unsigned(27 downto 0) := X"0000000";
    signal pb_count        : unsigned(23 downto 0) := X"000000";
    signal pb_d1           : std_logic := '1';
    signal pb_active       : std_logic := '0';
    signal pb_active_d1    : std_logic := '0';
    signal pb_toggle       : std_logic := '0';

begin
    pb_process : process(clk)
    begin
        if rising_edge(clk) then
            pb_d1 <= pb_in;   --synchronize
            --
            if (pb_d1 = '0') then
                pb_count <= DEBOUNCE;
            elsif (pb_count /= X"000000") then
                pb_count <= pb_count - 1;
            end if;
            --
            if (pb_d1 = '0') then
                pb_active <= '1';
            elsif (pb_count = X"000001") then
                pb_active <= '0';
            end if;
            -- rising edge detection and toggle
            pb_active_d1 <= pb_active;
            if ( pb_active = '1' AND pb_active_d1 = '0' ) then
                pb_toggle <= not pb_toggle;
            end if;
        end if;
    end process;

    led_counter : process(clk)
    begin
        if rising_edge(clk) then
            counter_led <= counter_led + 1;
        end if;
    end process;

    led <= pb_toggle AND counter_led(27);

end architecture Behavioral;
```

Fig. 2-12

1) The LED counter, sand the toggle gating of the MSB, and;

2) the debounce counter, pb_count.

Which leaves the new logic:

```vhdl
architecture Behavioral of push_button_2 is

    -- ~1 million
    constant DEBOUNCE      : unsigned(23 downto 0) := X"100000";

    signal counter_led     : unsigned(27 downto 0) := X"0000000";
    signal pb_count        : unsigned(23 downto 0) := X"000000";
    signal pb_d1           : std_logic := '1';
    signal pb_active       : std_logic := '0';
    signal pb_active_d1    : std_logic := '0';
    signal pb_toggle       : std_logic := '0';

begin
    pb_process : process(clk)
    begin
        if rising_edge(clk) then
            pb_d1 <= pb_in;  --synchronize
            --
            if (pb_d1 = '0') then
                pb_count <= DEBOUNCE;
            elsif (pb_count /= X"000000") then
                pb_count <= pb_count - 1;
            end if;
            --
            if (pb_d1 = '0') then
                pb_active <= '1';
            elsif (pb_count = X"000001") then
                pb_active <= '0';
            end if;
            -- rising edge detection and toggle
            pb_active_d1 <= pb_active;
            if ( pb_active = '1' AND pb_active_d1 = '0' ) then
                pb_toggle <= not pb_toggle;
            end if;
        end if;
    end process;

    led_counter : process(clk)
    begin
        if rising_edge(clk) then
            counter_led <= counter_led + 1;
        end if;
    end process;

    led <= pb_toggle AND counter_led(27);

end architecture Behavioral;
```

1 *2* *3*

Fig. 2-13

1) The new logic;

2) The set/reset pb_active latch; and
3) The combined rising-edge detection and the toggle.

Adjusted PB waveform

Perhaps the easiest way to show the operation of the rising-edge detection is with a waveform:

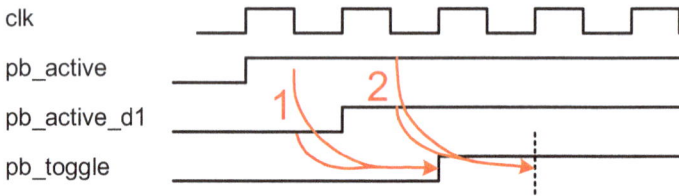

Fig. 2-14

1) pb_active is high, and the delayed pb_active_d1 is low, so pb_toggle inverts.
2) On the next clock, both are now high, and so pb_toggle is unchanged.

Chapter 3

Reflex Game

Now that we can light LEDs and take inputs from push-button switches, we'll use them both to implement a little game of reflexes. The idea is that the FPGA cues the user with an indication—a "span" signal—from the LED, and then waits a certain amount of time before blinking the LED again, a target to be hit. The trick is that the user must press the push-button at precisely the time the FPGA blinks the LED. If the user is successful, the LED winks an acknowledgment. The FPGA randomly changes the length of time that the "span" cue is lit. It then waits that same amount of time before blinking the "target." Thus, the user tries to match the time length of the "span" period with when the "target" blink occurs.

Here's a flow diagram:

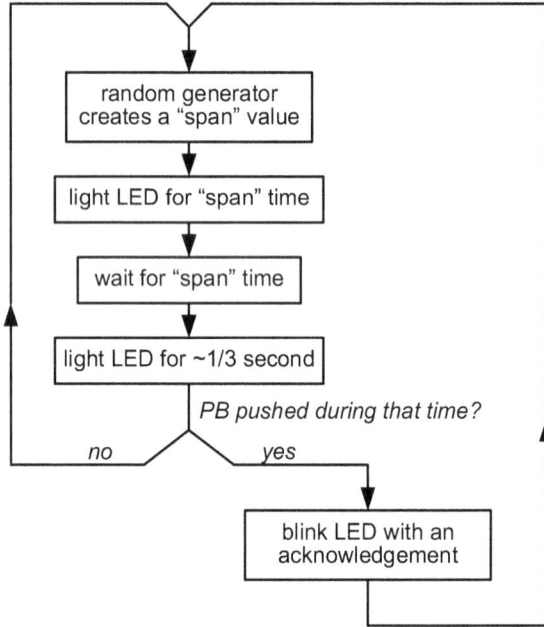

Fig. 3-1

And here's the associated state diagram:

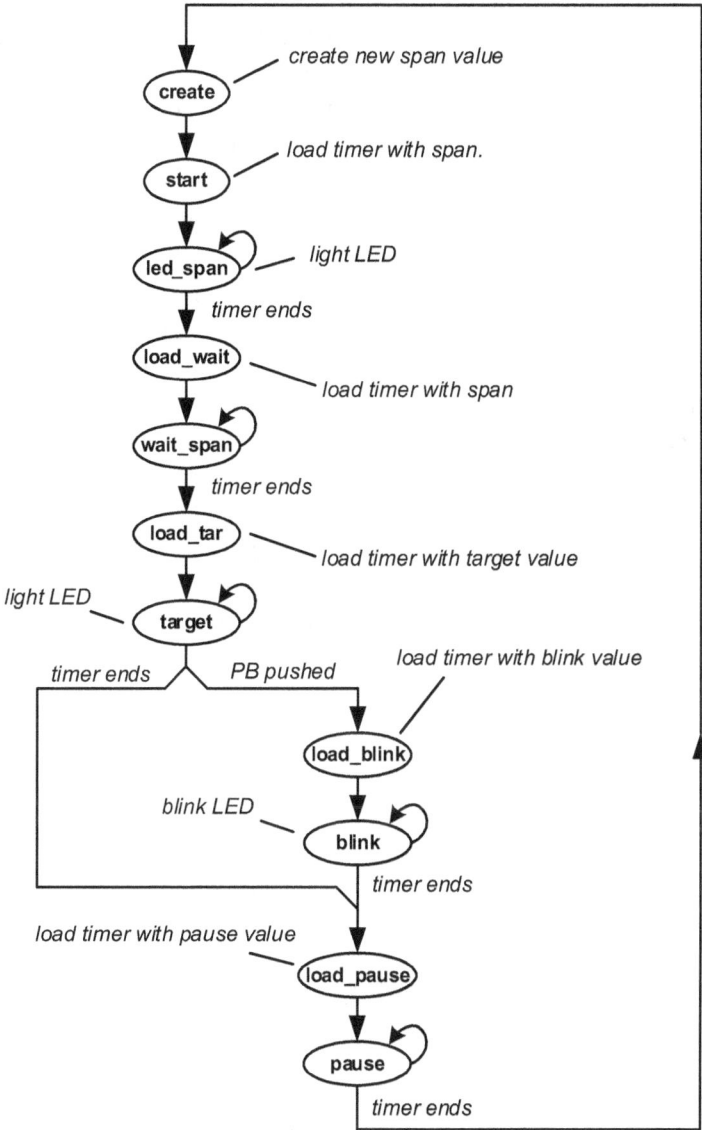

Fig. 3-2

State machines often have an idle state, waiting for the action to begin. Our operation, however, is ever-active, cycling between target offerings, so we'll begin at the "create" state, as though a previous target offering just completed. During the clock cycle of the "create" state (states, of course, operate on fixed clock periods) our logic creates a new pseudo-random span value. On the next clock cycle—the "start" state—we load a timer with this new span value. During the "led_span" state, the LED is lit—this is when the user is trying to assess the next target time. When the timer ends, the "load_wait" state again loads the timer with the same span value. However, during the subsequent "wait_span" state, the LED is not lit. This is the time that the user is waiting, ready to push the button when they guess that the span time is over. Once the timer again ends, the "load_tar" state loads the timer with the fixed target value (e.g., 1/3 second), and then the "target" state is the user's opportunity to push the button. The LED is lit during the "target" state, but 1/3 second is hardly enough for a finger to react in time—the user must anticipate the target.

Should the user be successful with a well-timed push, the timer is loaded with the fixed blink value (e.g., two seconds), and the LED is blinked during the "blink" state. The timer is then loaded with one last value during the "load_pause" state, and the "pause" state waits for that period of time before returning back to the "create" state to start the next cycle. This last "pause" period of time provides the user time (a few seconds) to prepare for the next cycle run.

If the user is not successful in timing the button push, the state machine simply skips the blinking LED acknowledge stages.

VHDL game code sections

Our VHDL code is now too extensive to view on one page of this book. Here, and going forward, we'll break the code into sections:

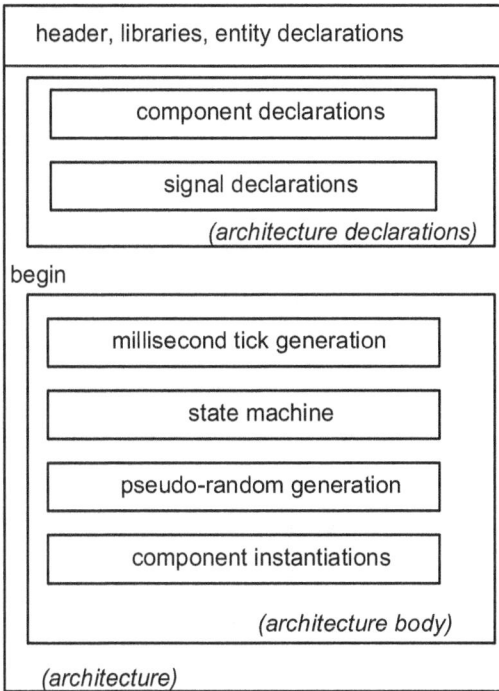

```
┌─────────────────────────────────────────────┐
│  header, libraries, entity declarations       │
├─────────────────────────────────────────────┤
│  ┌───────────────────────────────────────┐   │
│  │       component declarations           │   │
│  ├───────────────────────────────────────┤   │
│  │        signal declarations             │   │
│  └───────────────────────────────────────┘   │
│               (architecture declarations)     │
│  begin                                         │
│  ┌───────────────────────────────────────┐   │
│  │      millisecond tick generation        │   │
│  ├───────────────────────────────────────┤   │
│  │           state machine                 │   │
│  ├───────────────────────────────────────┤   │
│  │       pseudo-random generation          │   │
│  ├───────────────────────────────────────┤   │
│  │       component instantiations          │   │
│  └───────────────────────────────────────┘   │
│                    (architecture body)         │
│  (architecture)                                │
└─────────────────────────────────────────────┘
```

Fig. 3-3

The "header, libraries, and entity declarations" is the same as the previous design, in fact, exactly the same;

Within the architecture declarations section, in addition to the signal declarations we've seen previously, we now introduce the concept of VHDL components, and here we include the declarations of these (essentially simply the inputs and outputs);

We're breaking the architecture body into functional parts. Since our new design involves timing that spans periods of seconds, it's handy to use units of time as measured in milliseconds. We create these "millisecond ticks" in the first functional section. The next section is the state machine already shown. In order for our game to include random spans of time, we create a pseudo-random value in the next section ("pseudo" because it's not exactly random, but close enough for our purpose). And finally, we instantiate the functional modules that we declared above.

First game code section

This is the code for the header, libraries, and entity declarations:

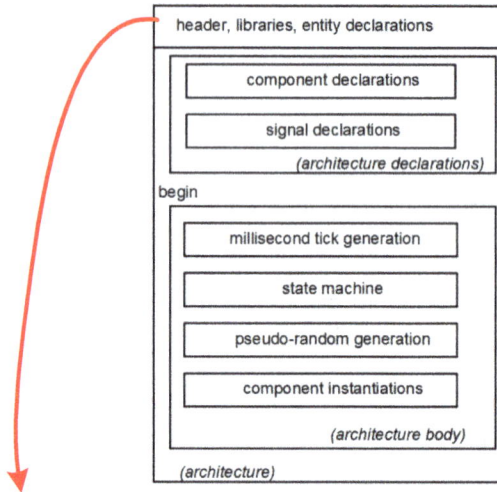

```
--  --------------------------------------------------------
--   Useful information goes here at the top.
--  --------------------------------------------------------

library IEEE;
use IEEE.STD_LOGIC_1164.all;
use IEEE.NUMERIC_STD.all;
use IEEE.STD_LOGIC_MISC.all;
use IEEE.STD_LOGIC_UNSIGNED.all;

entity target_game is
  port
    (
    clk             : in  std_logic; --50MHz
    pb_in           : in  std_logic; --low-active
    --
    led             : out std_logic
    );
end entity target_game;
```

Fig. 3-4

and you can see that this is identical to the I/Os of the previous design.

Game code component declarations

Here, you can see that the I/O declarations of our two component sub-modules are similar to that of the entity declaration above:

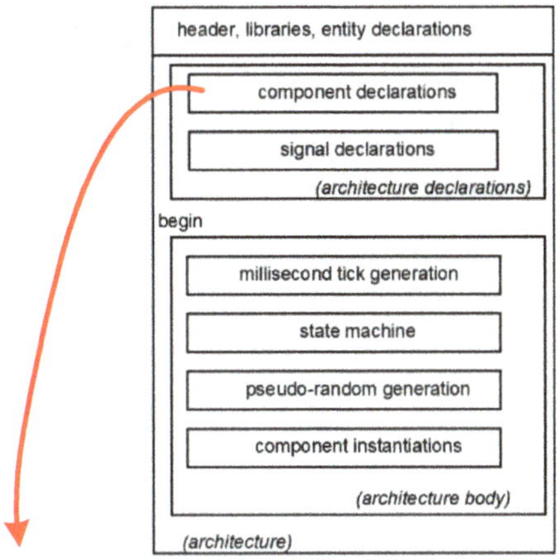

```vhdl
component debounce is
port
  (
     clk            : in   std_logic;
     pb_in          : in   std_logic; --push-button input
     ms_tick        : in   std_logic; --one millisecond indication
     pb_debnced     : out  std_logic  --push-button debounced output
  );
end component;

component timer is
port
  (
     clk            : in   std_logic;
     timer_load     : in   std_logic_vector(15 downto 0); --units of milliseconds
     timer_start    : in   std_logic; --begin the timer
     ms_tick        : in   std_logic; --one millisecond indication
     timer_blink    : out  std_logic; --one bit of the timer (toggling)
     timer_done     : out  std_logic  --indication the the timer has finished
  );
end component;
```

Fig. 3-5

Components are complete VHDL design files, and, in fact (as we'll see a bit later), these component declarations essentially become the entity declarations of those component entities (e.g., VHDL files). Component entities can include components of their own, allowing hierarchical structures. Indeed, our current design (target_game.vhd) could conceivably be instantiated in a higher-level hierarchy module, rendering it a sub-module in the hierarchy.

We'll dive into the details of the operation of these components later, but for now, we'll simply note that the "debounce" module serves the same function, and indeed is the same code, as the debounce operation of the previous design, while the "timer" component is the timer function referred to in the state machine summary above.

Game code signal declarations

Here's the signal declarations:

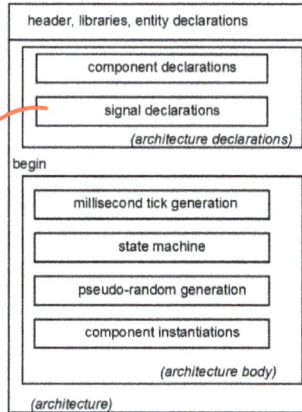

```
constant MS_LOAD      : unsigned(15 downto 0)              := X"C350"; --1ms at 50MHz
constant TAR_TIME     : std_logic_vector(15 downto 0) := X"0180"; --384 milliseconds
constant BLINK_TIME   : std_logic_vector(15 downto 0) := X"0800"; --2K (2 seconds)
constant PAUSE_TIME   : std_logic_vector(15 downto 0) := X"1000"; --4K (4 seconds)

signal ms_count       : unsigned(15 downto 0) := X"0001";
signal ms_tick        : std_logic := '0';
signal timer_load     : std_logic_vector(15 downto 0);
signal span           : std_logic_vector(15 downto 0);
signal timer_start    : std_logic := '0';
signal timer_done     : std_logic;
signal pb_debnced     : std_logic;
signal led_lcl        : std_logic;
signal timer_blink    : std_logic;
signal pseudo_sr      : std_logic_vector(1 to 10) := "1110100010";
signal pseudo_seg     : std_logic_vector(3 downto 0);

type state_type is ( create,
                     start,
                     led_span,
                     load_wait,
                     wait_span,
                     load_tar,
                     target,
                     load_blink,
                     blink,
                     load_pause,
                     pause
                   );

signal state : state_type := create;
```

Fig. 3-6

And we have a few new things to cover:

1

```
constant MS_LOAD      : unsigned(15 downto 0)        := X"C350"; --1ms at 50MHz
constant TAR_TIME     : std_logic_vector(15 downto 0) := X"0180"; --384 milliseconds
constant BLINK_TIME   : std_logic_vector(15 downto 0) := X"0800"; --2K (2 seconds)
constant PAUSE_TIME   : std_logic_vector(15 downto 0) := X"1000"; --4K (4 seconds)

signal ms_count       : unsigned(15 downto 0) := X"0001";
signal ms_tick        : std_logic := '0';
signal timer_load     : std_logic_vector(15 downto 0);
signal span           : std_logic_vector(15 downto 0);
signal timer_start    : std_logic := '0';
signal timer_done     : std_logic;
signal pb_debnced     : std_logic;
signal led_lcl        : std_logic;
signal timer_blink    : std_logic;
signal pseudo_sr      : std_logic_vector(1 to 10) := "1110100010";
signal pseudo_seg     : std_logic_vector(3 downto 0);

type state_type is ( create,
                     start,
                     led_span,
                     load_wait,
                     wait_span,
                     load_tar,
                     target,
                     load_blink,
                     blink,
                     load_pause,
                     pause
                   );

signal state : state_type := create;
```

2 **3** **4a** **4b**

Fig. 3-7

1) We were introduced to constants in the previous design, so this concept isn't new, but we can see here how convenient it is to assemble the various fixed timer values together. When initially developing the code, we might be changing—twiddling with—these values often (as I did when testing it), and it's nice to quickly find them;

2) Again, assigning an initial value is not new, and these, as we'll soon see, are functionally useful;

3) These signals are inputs to component modules in the following architecture body. In point of fact, virtually all FPGA devices power-up at a zero state unless set to an initial value of something else (as that ms_count signal), and so for that purpose these initial assignments are redundant. However, as we'll eventually see, we often must set initial values for the sake of simulation;

4a) VHDL includes enumerated types of signals. If you've used other programming languages, you are probably familiar with enumerated types, but if not, suffice it to say that enumeration allows you, the programmer, to create signals with your own set of values, usually labels. Here, we are making up a new type of signal that we're calling "state_type," and we're assigning to it a whole list of labels that you'll recognize as the state machine states. Note that the compiler will convert these labels into unique binary values, so using locally defined types is therefore merely a convenience (a very valuable convenience at that) for you, the programmer.

4b) And, once we've established our new "state_type" signal type, we use it to define the declared signal "state" (to which we assign an initial value of "create"). It is this signal, "state" that we will use in the logic coding of the architecture body below.

Game code architecture – millisecond tick

Moving on to the functional logic (the architecture body), we begin with the generation of millisecond timing "ticks," one clock-long flags occurring once each millisecond.

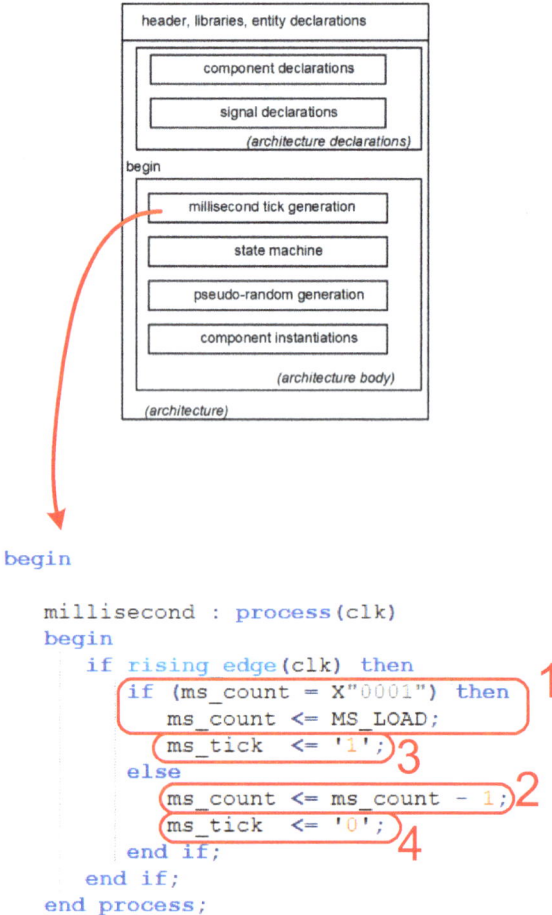

```
header, libraries, entity declarations
    component declarations
    signal declarations
                    (architecture declarations)
begin
    millisecond tick generation
    state machine
    pseudo-random generation
    component instantiations
                    (architecture body)
(architecture)
```

```
begin

    millisecond : process(clk)
    begin
        if rising_edge(clk) then
            if (ms_count = X"0001") then        1
                ms_count <= MS_LOAD;
                ms_tick  <= '1';               3
            else
                ms_count <= ms_count - 1;      2
                ms_tick  <= '0';               4
            end if;
        end if;
    end process;
```

Fig. 3-8

1) We begin by assuming that "ms_count" is a counter that is counting down. When it reaches one, it is reloaded with the MS_LOAD constant, a value that causes the resulting down-count to be one millisecond in length (X"C350" = 50,000; 20ns at 50MHz x 50,000 = one millisecond);

2) Otherwise, when the count isn't being reloaded, the counter down-counts here;

3) Each time the counter is reloaded, the "ms_tick" signal goes high for one clock …

4) … being reset on the next clock.

Game code state machine

Our state machine consists of two parts:

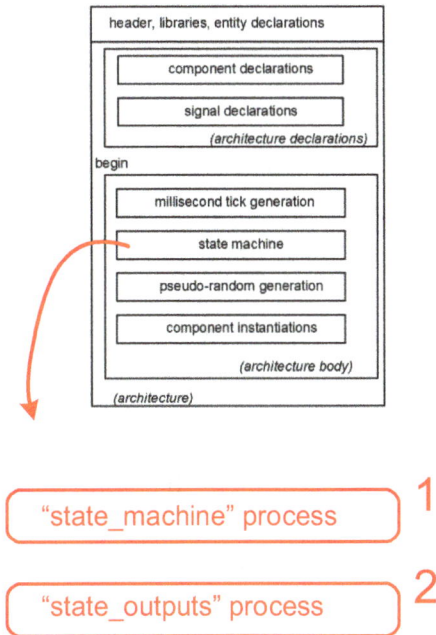

Fig. 3-9

1) The "state_machine" clocked process implements the pure state machine logic—the individual states, and the decisions about moving among them;

2) The "state_outputs" clocked process generates the signals associated with the different states, essentially the outputs of the state machine. You may have heard of the two basic configurations of state machine implementation: the Moore machine vs. the Mealy machine. These divisions, the fundamental approaches to constructing state machines, were important in the early days of limited logic resources, where every register was a valuable commodity. Now, though, with seemingly unlimited logic in modern FPGAs, the distinction is

more important to students of mathematical abstractions. (That said, as it happens, this state machine implements a Moore version, since each output is determined solely by the state condition, versus a combination of the state condition and inputs).

Here's the "state_machine" process:

```
state_machine : process(clk)
begin
   if rising_edge(clk) then
      case(state) is
         when create =>                                     state <= start;
         when start =>                                       state <= led_span;
         when led_span => if (timer_done = '1') then         state <= load_wait; end if;
         when load_wait =>                                   state <= wait_span;
         when wait_span => if (timer_done = '1') then        state <= load_tar; end if;
         when load_tar =>                                    state <= target;
         when target =>       if (pb_debnced = '1') then     state <= load_blink;
                              elsif (timer_done = '1') then  state <= load_pause;
                              end if;
         when load_blink =>                                  state <= blink;
         when blink =>        if (timer_done = '1') then     state <= load_pause; end if;
         when load_pause =>                                  state <= pause;
         when pause =>        if (timer_done = '1') then     state <= create; end if;
         when others => state <= create;
      end case;
   end if;
end process;
```

Fig. 3-10

And here's the breakdown.

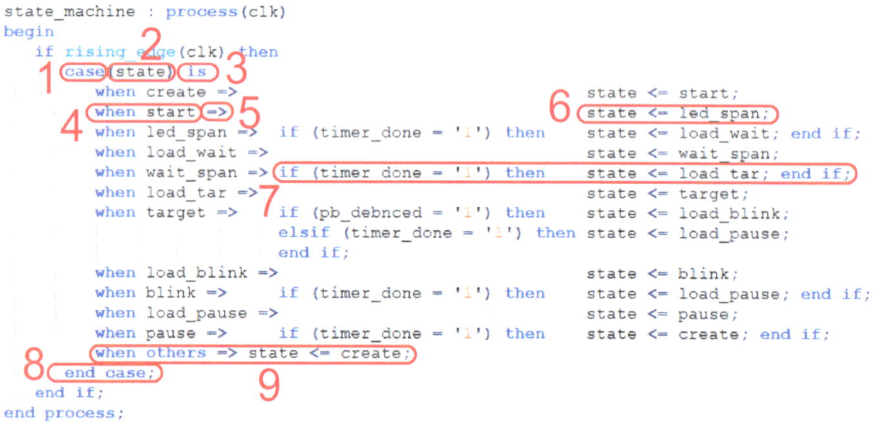

Fig. 3-11

1) We're using a VHDL "case" statement to implement the state machine. Those who've done any sort of other

programming will be familiar with these, but in effect, a case statement simply selects from among a list of possibilities;

2) In our case, the list is all the labels of the "state_type" we defined above, as contained in the "state" signal;

3) The "is" here is a VHDL key word, and must be included.

4) Each "when xxx" line represents a selection in the list. For example, "when" the "state" signal is "start", then this line is selected;

5) the "=>" combination is required, and simply indicates that the action to be taken follows;

6) In this case (for the "when start" line), the action is that the "led_span" type value is assigned to the "state" signal. Notice that on the next clock, since we've assigned "led_span" to "state", the case statement will select the next line ("when led_span"). In this way, the case statement steps through the transitions of the state machine;

7) The action taken in a selected line can be a conditional statement, and this is how we implement decisions in the state machine. Notice that we don't need to define an "else" condition. In other words, we don't need to tell the compiler to stay on "wait_span" if "timer_done" is not a one;

8) The case list ends with the VHDL keywords "end case";

9) Often times, the compiler will require you to define a default action—what the compiler should do if "state" somehow ends up with a value not included in the list. This is the "others" VHDL keyword. In this case, we're telling the compiler to default to the "create" state.

Game code state machine outputs

The previous section of code does nothing other than implement the states of the state machine. The following code now implements the outputs of the state machine—i.e., the results.

```vhdl
state_outputs : process(clk)
begin
   if rising_edge(clk) then
      timer_start <= '0';
      led_lcl     <= '0';
      --
      case(state) is
         when start =>        timer_load  <= span;
                              timer_start <= '1';
         when led_span =>     led_lcl     <= '1';
         when load_wait =>    timer_load  <= span;
                              timer_start <= '1';
         when load_tar =>     timer_load  <= TAR_TIME;
                              timer_start <= '1';
         when target   =>     led_lcl     <= '1';
         when load_blink =>   timer_load  <= BLINK_TIME;
                              timer_start <= '1';
         when blink =>        led_lcl     <= timer_blink;
         when load_pause =>   timer_load  <= PAUSE_TIME;
                              timer_start <= '1';
         when others => null;
      end case;
   end if;
end process;

led <= led_lcl;
```

Fig. 3-12

Here we have another case statement:

```vhdl
state_outputs : process(clk)
begin
   if rising_edge(clk) then
      timer_start <= '0';
      led_lcl    <= '0';
      --
      case(state) is                1              3
         when start =>        timer_load  <= span;
                              timer_start <= '1';
         when led_span =>     led_lcl     <= '1';
         when load_wait =>    timer_load  <= span;
                              timer_start <= '1';
    2    when load_tar =>     timer_load  <= TAR_TIME;
                              timer_start <= '1';
         when target   =>     led_lcl     <= '1';
         when load_blink =>   timer_load  <= BLINK_TIME;
                              timer_start <= '1';
         when blink =>        led_lcl     <= timer_blink;
         when load_pause =>   timer_load  <= PAUSE_TIME;
                              timer_start <= '1';
         when others => null;          4
      end case;
   end if;
end process;

led <= led_lcl;  5
```

Fig. 3-13

In fact, it uses nearly the same case statement selection list as the previous one.

1) The selection list is again the "state" signal,

2) ... except that the selection list is a subset of the entire set of "state" type labels. Not all of the states are associated with signals that become active (the "wait_span" and "pause" states produce no active signals);

3) Note that selected cases can have multiple "actions", e.g., both loading and starting the timer;

4) While the compiler may require a default "others" case selection, we don't necessarily need to give it a specific selection to perform. When we use "null", we're basically telling the compiler to chill, do nothing. Since this case statement is not "reactive", meaning the sequence of states is not determined here, establishing a default is somewhat superfluous;

5) It may seem strange that we created the "led_lcl" signal just to then directly assign it to the output "led". This is a

common practice, since VHDL does not allow us to use outputs signals within the architecture body. In this case, led_lcl could be used, for example, during initial debug to access the led signal locally in diagnostic exercises (as I did—the "_lcl" addition to the name is a handy way to remember that this is a "local" signal).

Lastly, we have a couple of zero assignments that we need to understand:

```
state_outputs : process(clk)
begin
  if rising edge(clk) then
      timer_start <= '0';
      led_lcl     <= '0';            1
      --
      case(state) is
         when start =>      timer_load  <= span;        3
                            timer_start <= '1';
         when led_span =>   led_lcl     <= '1';
         when load_wait =>  timer_load  <= span;
                            timer_start <= '1';
         when load_tar =>   timer_load  <= TAR_TIME;    2
                            timer_start <= '1';
         when target   =>   led_lcl     <= '1';
         when load_blink => timer_load  <= BLINK_TIME;
                            timer_start <= '1';
         when blink =>      led_lcl     <= timer_blink;
         when load_pause => timer_load  <= PAUSE_TIME;
                            timer_start <= '1';
         when others => null;
      end case;
  end if;
end process;
```

Fig. 3-14

1) We assign zeros to two signals, "timer_last" and "led_lcl", prior to the case statement …

2) … where we then set one or the other when certain case selections are made. In order to explain what's going on, we need to understand the concept of concurrent versus sequential operation. Assignments made outside of a process, at the level of the architecture body, are done concurrently, meaning that they all happen simultaneously. It is for this reason that if you try to make more than one assignment to the same signal at the

architecture body level, the compiler will error, since it wouldn't know which assignment to make. Assignments inside a process, however, are performed sequentially, meaning that the compiler starts at the beginning and works down. If more than one assignment is encountered, then the later ones override the earlier. Thus, in this process, at each rising clock edge the two signals timer_start and led_lcl always begin as zero. The zero is replaced by a one if a case selection is made to an entry that sets one of the signals to one. On the next rising clock edge, that signal is then set back to zero (unless the next case selection again sets it to one). In a clocked process, like we've been using, this sequential operation is more or less hidden. To the rest of the design outside the process, the process is active for just an instant in time at each clock edge. Whatever sequential operations occurred between the clock edges is done behind the curtain, so to speak;

3) There are signals that are not set to zero prior to the case statement, and they remain as set until reassigned. These signals are thus implemented as latches. For timer_start and led_lcl it is essential that they be returned to zero, since they cause action (as we'll see). In the case of timer_load, on the other hand, its value doesn't matter during clock cycles where it's not used.

Comparing the earlier state diagram to these two processes you can see how together they implement the state machine. It is possible to collapse the two processes together, adding the second process output signal assignments to the first process case statement selections, and some designers do this, but I recommend keeping them separate, as it makes the state machine operation much easier to follow.

Game code pseudo-random generator

The next functional section is the pseudo-random generator.

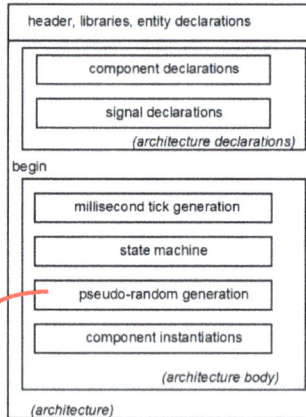

```
pseudo_random : process(clk)
begin
   if rising_edge(clk) then
      if (state = create) then
                        ------------ bit 1 ------------     -- bits 2-10 ----
         pseudo_sr <= (pseudo_sr(7) XOR pseudo_sr(10)) & pseudo_sr(1 to 9);
      end if;
   end if;
end process;

pseudo_seg <= pseudo_sr(1 to 4);

span <= X"0200" when pseudo_seg = X"0" else
        "000" & pseudo_seg & '0' & X"00";
```

Fig. 3-15

And here's the breakdown:

fig. 3-16

1) As we saw from the state machine earlier, a new pseudo-random ("almost random") value is created during the "create" state with each cycle of the game;

2) This is the actual generation of the pseudo-random value, and we'll look at how this works shortly. For now, just assume that a new ten-bit "pseudo_sr" value is created each "create" state;

3) We only need four bits of the pseudo-random value for our four-bit "pseudo_seg" segment, and since a new pseudo-random value is created with each generation, we could take any four bits out of the ten, and I chose bits 1-4;

4) Here we see a new symbol used, "&". The ampersand (this symbol) indicates concatenation. For example, if I have a two-bit signal "first" that equals binary "10", and another two-bit signal "second" that equals binary "01", then "second & first" would be a four-bit value of "0110". On the other hand, "first & second" would be "1001". The "span" signal is 16 bits, but our timer counter only needs thirteen bits for the full range of the game, thus the three "000" zero MS bits. So we can see that our pseudo-random four-bit segment comprises the MS four bits of a thirteen-bit value used by our timer.

5) And here we encounter a new type of conditional statement. Whereas the "if/then/else" type is used inside processes, this "when/else" is used directly in the architecture body (outside a process). The operation is more or less self-explanatory, "when" pseudo_seg is zero, X"0200" is assigned to "span", otherwise ("else") it's the pseudo_seg based value we just analyzed. The reason for using this conditional selection is that we don't want an occasional zero pseudo_seg to result in a zero "span" (a zero length time in the game). Instead, when pseudo_seg is zero, we substitute a minimum value for it (as though pseudo_seg were "0001" instead of "0000"). Note that by definition (being outside process statements), "when/else" assignments are continuous, i.e., implemented as gate logic instead of registers (as in clocked process statements).

Now we'll step back and look at the pseudo-random generation line. But first, we look at the signal declaration of "pseudo_sr".

```
signal pseudo_sr      : std_logic_vector(1 to 10) := "1110100010";
```

Fig. 3-17

This is different than the "std_logic_vector(9 downto 0)" declaration for a ten-bit value that we've already seen. This one is unusual in two ways: first, instead of 0 to 9, this one is 1 to 10, and instead of the largest bit "downto" the smallest bit, this declaration is the smallest bit up "to" the largest. In use, as long as we don't try to identify individual bits—as long as we use the entire vector—the two declarations result in the same operation. Once we try to tag—to pick out—any comprising bits, though, we need to be cognizant of the form of the declaration.

Here's the functional operation of pseudo-random generation statement:

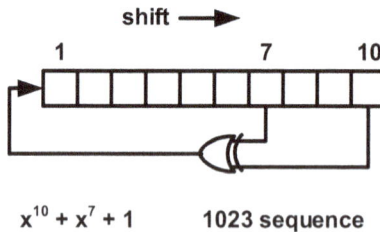

$$x^{10} + x^7 + 1 \qquad 1023 \text{ sequence}$$

Fig. 3-18

This illustrates a classic pseudo-random generator, where bits of a shift register are XOR'd and fed back to the input. The mathematical theory for this is far beyond the scope of this book, and if you'd like to learn more, Wikipedia has (currently) an excellent entry. Comparing this diagram to the VHDL statement line, you can see why we declared the signal as 1-to-10, and in ascending order—this makes the correlation with the diagram clearer. Study the VHDL statement line closely, and you'll see how it implements a shift (right) register, where with each update (when state = create), the MS bit is replaced with the XOR'd result, and the MS nine bits (1 to 9) are moved to the LS nine bit position.

Game code component instantiations

We finally now come to the component instantiations, where we actually use them functionally.

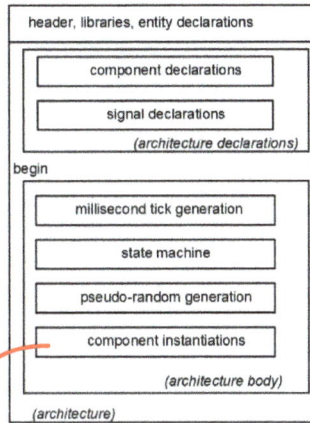

```
debounce_ins : debounce
port map
  (
    clk            => clk,          --in  std_logic;
    pb_in          => pb_in,        --in  std_logic;
    ms_tick        => ms_tick,      --in  std_logic;
    pb_debnced     => pb_debnced    --out std_logic
  );

timer_inst : timer
port map
  (
    clk            => clk,          --in  std_logic;
    timer_load     => timer_load,   --in  std_logic_vector(15 downto 0);
    timer_start    => timer_start,  --in  std_logic;
    ms_tick        => ms_tick,      --in  std_logic;
    timer_blink    => timer_blink,  --out std_logic;
    timer_done     => timer_done    --out std_logic
  );
```

Fig. 3-19

Looking at the form:

Fig. 3-20

1) Each component instantiation starts with a unique name that we make up. In this case, these are "debounce_ins" and "timer_inst", followed, via a colon, by the name of the component as declared above—in this case, "debounce" and "timer" (I added "_ins" and "inst" to the end of the component names as abbreviations of "instantiation", but the names could have been "david" and "alice" with equal results).

2) The instantiation and component names are followed by the key words "port map".

Continuing …

```
debounce_ins : debounce
port map   1    2    3
  (
    clk         => clk,          --in  std_logic;
    pb_in       => pb_in,        --in  std_logic;
    ms_tick     => ms_tick,      --in  std_logic;
    pb_debnced  => pb_debnced    --out std_logic
  );

timer_inst : timer
port map
  (
    clk          => clk,          --in  std_logic;
    timer_load   => timer_load,   --in  std_logic_vector(15 downto 0);
    timer_start  => timer_start,  --in  std_logic;
    ms_tick      => ms_tick,      --in  std_logic;
    timer_blink  => timer_blink,  --out std_logic;
    timer_done   => timer_done    --out std_logic
  );
```

Fig. 3-21

Within parenthesis after the "port map" key words, we assign an architecture signal to each I/O of the component (thus we "map" local signals to the component).

1) These are the component I/O signal names as defined in the component declaration above;

2) The "=>" combination is a key word set that constitutes a component/local signal connection;

3) These are the local architecture signals. Note that all these signals are declared above.

Continuing more …

```
debounce_ins : debounce
port map
  (
    clk            => clk,        --in  std_logic;
    pb_in          => pb_in,      --in  std_logic;
    ms_tick        => ms_tick,    --in  std_logic;
    pb_debnced     => pb_debnced  --out std_logic
  );
timer_inst : timer
port map
  (
    clk            => clk,          --in  std_logic;
    timer_load     => timer_load,   --in  std_logic_vector(15 downto 0);
    timer_start    => timer_start,  --in  std_logic;
    ms_tick        => ms_tick,      --in  std_logic;
    timer_blink    => timer_blink,  --out std_logic;
    timer_done     => timer_done    --out std_logic
  );
```

Fig. 3-22

1) Each local architecture signal is followed by a comma,
2) … except the last one.
3) The ending parenthesis is followed by a semi-colon.

Game code debounce component

As we've seen, the components themselves are separate VHDL entities, separate files. Here's the debounce component:

```vhdl
-- ----------------------------------------------------------
--    Useful information goes here at the top.
-- ----------------------------------------------------------

library IEEE;
use IEEE.STD_LOGIC_1164.all;
use IEEE.NUMERIC_STD.all;
use IEEE.STD_LOGIC_MISC.all;
use IEEE.STD_LOGIC_UNSIGNED.all;

entity debounce is
  port
    (
      clk            : in  std_logic;
      pb_in          : in  std_logic;
      ms_tick        : in  std_logic;
      pb_debnced     : out std_logic
    );
end entity debounce;

architecture Behavioral of debounce is

    -- ~1 million
    constant DEBOUNCE_TIME : unsigned(11 downto 0) := X"3E8";

    signal pb_d1          : std_logic;
    signal pb_active      : std_logic;
    signal pb_active_d1   : std_logic;
    signal pb_count       : unsigned(11 downto 0) := X"000";
begin

    pb_process : process(clk)
    begin
      if rising_edge(clk) then
        pb_d1 <= pb_in;  --synchronize
        --
        if (pb_d1 = '0') then
           pb_count <= DEBOUNCE_TIME;
        elsif (ms_tick = '1') then
           if (pb_count /= X"000") then
              pb_count <= pb_count - 1;
           end if;
           --
        end if;
        --
        if (pb_d1 = '0') then
           pb_active <= '1';
        elsif (pb_count = X"001") then
           pb_active <= '0';
        end if;
        -- rising edge detection and toggle
        pb_active_d1 <= pb_active;
        if ( pb_active = '1' AND pb_active_d1 = '0' ) then
           pb_debnced <= '1';
        else
           pb_debnced <= '0';
        end if;
      end if;
    end process;

end architecture Behavioral;
```

Fig. 3-23

Blaine C. Readler

The basic operation is the same as the debounce portion of the previous design, except for the addition of the millisecond tick, "ms_tick".

```vhdl
--  ------------------------------------------------------------
--     Useful information goes here at the top.
--  ------------------------------------------------------------

library IEEE;
use IEEE.STD_LOGIC_1164.all;
use IEEE.NUMERIC_STD.all;
use IEEE.STD_LOGIC_MISC.all;
use IEEE.STD_LOGIC_UNSIGNED.all;

entity debounce is
  port
    (
      clk                : in  std_logic;
      pb_in              : in  std_logic;
      ms_tick            : in  std_logic;
      pb_debnced         : out std_logic
    );
end entity debounce;

architecture Behavioral of debounce is                1
   -- ~1 million
   constant DEBOUNCE_TIME : unsigned(11 downto 0) := X"3E8";

   signal pb_d1           : std_logic;
   signal pb_active       : std_logic;
   signal pb_active_d1    : std_logic;                 2
   signal pb_count        : unsigned(11 downto 0) := X"000";
```

Fig. 3-24

Using the millisecond tick allows our counter to be much smaller—instead of counting 50MHz clocks for the debounce period, it can count milliseconds, a one-thousand-fold decrease in counts.

1) Thus, instead of a debounce time of one second consisting of a 24-bit value of X"100000" for the previous design, the new load value is just twelve bits at X"3E8";

2) And, of course, the counter is also just twelve bits instead of twenty-four.

Here's the body of the debounce component:

56

```
begin

    pb_process : process(clk)
    begin
        if rising edge(clk) then
            pb_d1 <= pb_in;   --synchronize
            --
            if (pb_d1 = '0') then
                pb_count <= DEBOUNCE_TIME;
            elsif (ms_tick = '1') then
                if (pb_count /= X"000") then
                    pb_count <= pb_count - 1;
                end if;
                --
            end if;
            --
            if (pb_d1 = '0') then
                pb_active <= '1';
            elsif (pb_count = X"001") then
                pb_active <= '0';
            end if;
            -- rising edge detection and toggle
            pb_active_d1 <= pb_active;
            if ( pb_active = '1' AND pb_active_d1 = '0' ) then
                pb_debnced <= '1';
            else
                pb_debnced <= '0';
            end if;
        end if;
    end process;

end architecture Behavioral;
```

Fig.
3-25

1) These sections are the same as the previous design;
2) while this section is the same, except that the count decrement is now qualified by the millisecond tick, "ms_tick", thus decrementing once each millisecond.

Game code timer component

The "timer" component is new for this design, serving as a general timer in support of the state machine.

```
--  ----------------------------------------------------------
--    Useful information goes here at the top.
--  ----------------------------------------------------------

library IEEE;
use IEEE.STD_LOGIC_1164.all;
use IEEE.NUMERIC_STD.all;
use IEEE.STD_LOGIC_MISC.all;
use IEEE.STD_LOGIC_UNSIGNED.all;

entity timer is
  port
    (
       clk             : in  std_logic;
       timer_load      : in  std_logic_vector(15 downto 0); --ms units
       timer_start     : in  std_logic;
       ms_tick         : in  std_logic;
       timer_blink     : out std_logic;
       timer_done      : out std_logic
    );
end entity timer;

architecture Behavioral of timer is


    signal timer_count    : unsigned(15 downto 0) := X"0000";

begin

    timer_process : process(clk)
    begin
      if rising_edge(clk) then
         if (timer_start = '1') then
            timer_count <= unsigned(timer_load);
         elsif (ms_tick = '1') then
            if (timer_count /= "0000") then
               timer_count <= timer_count - 1;
            end if;
         end if;
         --
         if (     ms_tick = '1'
              and timer_count = "0001"
            ) then
            timer_done <= '1';
         else
            timer_done <= '0';
         end if;
      end if;
    end process;

    timer_blink <= timer_count(8); --blinks every 1/4 second

end architecture Behavioral;
```

Fig. 3-26

The only internal signal for this component is the timer counter itself.

```vhdl
-- --------------------------------------------------------------
--    Useful information goes here at the top.
-- --------------------------------------------------------------

library IEEE;
use IEEE.STD_LOGIC_1164.all;
use IEEE.NUMERIC_STD.all;
use IEEE.STD_LOGIC_MISC.all;
use IEEE.STD_LOGIC_UNSIGNED.all;

entity timer is
  port
    (
        clk             : in  std_logic;
        timer_load      : in  std_logic_vector(15 downto 0); --ms units
        timer_start     : in  std_logic;
        ms_tick         : in  std_logic;
        timer_blink     : out std_logic;
        timer_done      : out std_logic
    );
end entity timer;

architecture Behavioral of timer is


    signal timer_count      : unsigned(15 downto 0) := X"0000";

begin
```

Fig. 3-27

Here's the timer component body:

```vhdl
timer_process : process(clk)
begin
   if rising_edge(clk) then
      if (timer_start = '1') then
         timer_count <= unsigned(timer_load);
      elsif (ms_tick = '1') then
         if (timer_count /= "0000") then
            timer_count <= timer_count - 1;
         end if;
      end if;
      --
      if (     ms_tick = '1'
         and timer_count = "0001"
         ) then
         timer_done <= '1';
      else
         timer_done <= '0';
      end if;
   end if;
end process;

timer_blink <= timer_count(9); --blinks every 1/4 second

end architecture Behavioral;
```

Fig. 3-28

1) The "timer_start" input loads the counter with the provided "timer_load" value;

2) Since "timer_count" was declared as an unsigned value (so that it could count), we need convert "timer_load" to unsigned as well, and we can do this by simply declaring it so;

3) "timer_start" loads the counter immediately, otherwise it waits for the next millisecond boundary ("ms_tick") to decrement, freezing when it reaches zero;

4) When the timer passes through a value of one (having now counted the time span of "timer_load"), the "timer_done" output is set;

5) We must qualify the setting of "timer_done" with "ms_tick", otherwise "timer_done" would be active for a full millisecond, when we want it to be high for just one clock period (the one dictated by "ms_tick");

6) We take advantage of the fact that the various bits of the timer counter toggle at regular rates to produce the blinking LED signal. We choose bit 8 since that one toggles every 512 counts, and since the timer counter counts every millisecond, that means that the signal toggles every 512 milliseconds, or 1/4 second.

Blaine C. Readler

Chapter 4

Simulation

Software programs flow ethereally through a computer's core, directing the operation of memory and peripherals one step at a time, sharing or replacing other programs for awhile, and then evaporating away to be replaced by other programs performing their sequence of specific steps. Compiled FPGA code, on the other hand, performs what otherwise would be implemented with fixed logic and wires—millions of gates and wires all functioning together all the time. Unlike software, whose contact with the outside universe is routed through the computer hardware surrounding it, FPGAs can interface directly with the real world.

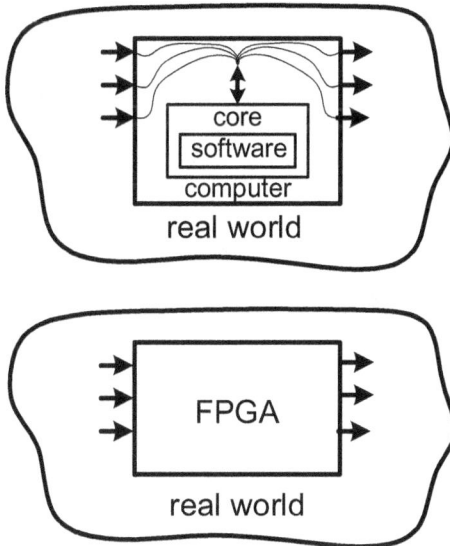

Fig. 4-1

You may have already tried the last simple target game program on an actual FPGA development board such as offered by various companies like Digilent.

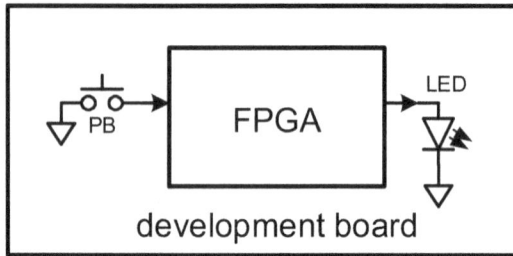

Fig. 4-2

If not, then the operation is somewhat abstract, imagined in your head. An intermediate approach, involving just ethereal software running on your computer, would be to simulate the design. Most FPGA vendors offer a basic simulation tool along with their compiling software, usually free for the smaller FPGAs that would be used in the course of this book. As I write this, Intel/Altera offers a free starter version of the very popular Modelsim simulation software when downloading their free version of their Quartus II compiler tool, while Xilinx provides built-in simulation (with a look and feel similar to Modelsim) with the free version of their Vivado compiler. Both of these free compiler and simulation versions have the capacity to handle all the material covered in this text.

Simulation testbench concept

In simulation, we emulate the operation of external inputs, in this case the push-button switch, and we monitor the outputs, here the LED, usually on a waveform viewer:

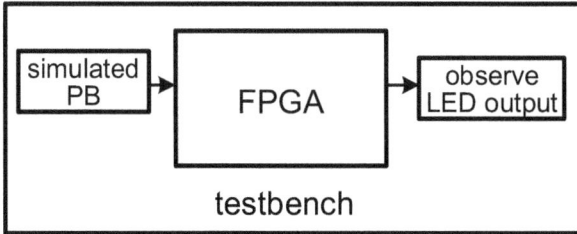

Fig. 4-3

Instead of loading our compiled code into an actual FPGA and interfacing with physical push-buttons and LEDs, we can simulate the real world with something we call a testbench, which is nothing more than another VHDL file that instantiates our top-level entity as though our design is a sub-module, which in this case it is.

Here's our testbench for the target game design:

```vhdl
--  Push Button Testbench
--------------------------------------------------------------------
library IEEE;
use IEEE.STD_LOGIC_1164.all;
use IEEE.NUMERIC_STD.all;
use IEEE.STD_LOGIC_MISC.all;
use IEEE.STD_LOGIC_UNSIGNED.all;

entity target_game_tb is
end entity target_game_tb;

architecture Testbench of target_game_tb is

    component target_game_sim is
      generic
      (  simulation : boolean
      );
      port
      (
        clk             : in  std_logic; --50MHz
        pb_in           : in  std_logic; --low-active
        --
        led             : out std_logic
      );
    end component;

    signal clk            : std_logic := '0';
    signal pb_in          : std_logic := '0';
    signal tb_counter     : unsigned(15 downto 0) := X"0000";
    signal led            : std_logic;

begin

    ------  generate clock -----------

    clk  <= '0' after 10 ns when clk = '1' else    -- 50MHz = 20ns
            '1' after 10 ns;

    ------  Push-Button generation

    p_button : process(clk)
    begin
      if rising_edge(clk) then
          tb_counter <= tb_counter + 1;
          --
          case(tb_counter) is
            when X"000D" => pb_in <= '0';
            when X"000F" => pb_in <= '1';
            when X"0028" => pb_in <= '0';
            when X"002B" => pb_in <= '1';
            when others => null;
          end case;
      end if;
    end process;

    -----  design instantiation

    target_game_sim_inst : target_game_sim
      generic map
      ( simulation => true
      )
      port map
      (
        clk             => clk,   --in  std_logic; --50MHz
        pb_in           => pb_in, --in  std_logic; --low-active
        --
        led             => led    --out std_logic
      );

end architecture Testbench;
```

Fig. 4-4

This should look somewhat familiar, as it has the same form as any other VHDL file.

Simulation entity/architecture declarations

We'll look at the entity and architecture declarations first.

```vhdl
--   Push Button Testbench
---------------------------------------------------------------
library IEEE;
use IEEE.STD_LOGIC_1164.all;
use IEEE.NUMERIC_STD.all;
use IEEE.STD_LOGIC_MISC.all;
use IEEE.STD_LOGIC_UNSIGNED.all;

entity target_game_tb is
end entity target_game_tb;

architecture Testbench of target_game_tb is

  component target_game_sim is
    generic
      ( simulation : boolean
      );
    port
      (
      clk             : in  std_logic; --50MHz
      pb_in           : in  std_logic; --low-active
      --
      led             : out std_logic
      );
  end component;

  signal clk         : std_logic := '0';
  signal pb_in       : std_logic := '0';
  signal tb_counter  : unsigned(15 downto 0) := X"0000";
  signal led         : std_logic;

begin

------ generate clock -----------

clk  <= '0' after 10 ns when clk = '1' else    -- 50MHz = 20ns
        '1' after 10 ns;

------  Push-Button generation

  p_button : process(clk)
  begin
    if rising_edge(clk) then
      tb_counter <= tb_counter + 1;
      --
      case(tb_counter) is
        when X"000D" => pb_in <= '0';
        when X"000F" => pb_in <= '1';
        when X"0028" => pb_in <= '0';
        when X"0029" => pb_in <= '1';
        when others => null;
      end case;
    end if;
  end process;

----- design instantiation

  target_game_sim_inst : target_game_sim
    generic map
      ( simulation => true
      )
    port map
      (
      clk             => clk,   --in  std_logic; --50MHz
      pb_in           => pb_in, --in  std_logic; --low-active
      --
      led             => led    --out std_logic
      );

end architecture Testbench;
```

Fig. 4-5

Here they are in detail:

```
--   Push Button Testbench
----------------------------------------------------------------
library IEEE;
use IEEE.STD_LOGIC_1164.all;
use IEEE.NUMERIC_STD.all;
use IEEE.STD_LOGIC_MISC.all;
use IEEE.STD_LOGIC_UNSIGNED.all;

entity target_game_tb is                    1
end entity target_game_tb;

architecture  Testbench  of target_game_tb is

  component target_game_sim is              3
    generic
    ( simulation : boolean                  4
    );
    port
    (
      clk              : in  std_logic; --50MHz
      pb_in            : in  std_logic; --low-active
      --
      led              : out std_logic
    );
  end component;                                        5

  signal clk          : std_logic := '0';
  signal pb_in        : std_logic := '0';
  signal tb_counter   : unsigned(15 downto 0) := X"0000";
  signal led          : std_logic;

begin
```

Fig. 4-6

1) The testbench entity has no I/O. This makes sense, since it essentially constitutes the entire universe outside our simulated design;

2) I'm naming the architecture as "Testbench" simply because that makes sense as well;

3) The only component in this testbench is our simulated design. Other testbenches might include additional components as well, used in support of the testbench architecture operation;

4) We introduce a new component structure, the "generic", as well as a new signal type, "Boolean", both of which we'll get to shortly;

5) These are signals used by the testbench architecture in the generation of the simulated design inputs (in our case, just the clock and push-button input).

Simulation architecture/generics

We move on to the architecture itself:

```vhdl
--   Push Button Testbench
--------------------------------------------------------
library IEEE;
use IEEE.STD_LOGIC_1164.all;
use IEEE.NUMERIC_STD.all;
use IEEE.STD_LOGIC_MISC.all;
use IEEE.STD_LOGIC_UNSIGNED.all;

entity target_game_tb is
end entity target_game_tb;

architecture Testbench of target_game_tb is

    component target_game_sim is
      generic
        ( simulation : boolean
        );
      port
        (
         clk            : in  std_logic; --50MHz
         pb_in          : in  std_logic; --low-active
         --
         led            : out std_logic
        );
    end component;

    signal clk            : std_logic := '0';
    signal pb_in          : std_logic := '0';
    signal tb_counter     : unsigned(15 downto 0) := X"0000";
    signal led            : std_logic;

begin

    ------   generate clock ------------

    clk  <= '0' after 10 ns when clk = '1' else    -- 50MHz = 20ns
            '1' after 10 ns;

    ------   Push-Button generation

    p_button : process(clk)
    begin
       if rising_edge(clk) then
          tb_counter <= tb_counter + 1;
          --
          case(tb_counter) is
             when X"000D" => pb_in <= '0';
             when X"000F" => pb_in <= '1';
             when X"0028" => pb_in <= '0';
             when X"002B" => pb_in <= '1';
             when others => null;
          end case;
       end if;
    end process;

    ------ design instantiation

    target_game_sim_inst : target_game_sim
      generic map
        ( simulation => true
        )
      port map
        (
         clk            => clk,    --in  std_logic; --50MHz
         pb_in          => pb_in,  --in  std_logic; --low-active
         --
         led            => led     --out std_logic
        );

end architecture Testbench;
```

Fig. 4-7

With the details:

```
begin
------    generate clock ------------                              1

clk  <= '0' after 10 ns when clk = '1' else    -- 50MHz = 20ns
           '1' after 10 ns;

------    Push-Button generation                   2

p_button : process(clk)
begin
   if rising_edge(clk) then
       tb_counter <= tb_counter + 1;
       --
       case(tb_counter) is
           when X"000D" => pb_in <= '0';
           when X"000F" => pb_in <= '1';
           when X"0028" => pb_in <= '0';
           when X"002B" => pb_in <= '1';
           when others => null;
       end case;
   end if;
end process;

----- design instantiation                              3

target_game_sim_inst : target_game_sim
   generic map
   ( simulation => true )  4
   )
   port map
   (
       clk              => clk,   --in  std_logic; --50MHz
       pb_in            => pb_in, --in  std_logic; --low-active
       --
       led              => led    --out std_logic
   );

end architecture Testbench;
```

Fig. 4-8

1) The first signal we generate is the clock. Reading the lines of code, we can see that it essentially describes a toggling operation, which is, of course, exactly what a clock is. This form, (signal state) "after" (time), is used for gate level simulation (we are simulating at the RTL—"Register Transfer" Level), but is useful when creating a clock, which we can then use for our register creations. Here, since we're defining the clock as high and low for 10 nanoseconds each, the overall period is 20 nanoseconds, or 50MHz;

2) With this process, we generate the push-button input for our simulated design. We first create a continually incrementing counter (tb_counter), and use a case statement to turn the push-button on and off at specific times. You can see that we're simulating two pushes of

the button (low-active, remember). As we'll soon show, the first push misses the target, and the second hits it;

　　3) Finally, we have the instantiation of our simulated design.

　　4) We've added this generic to the original design. Generics are like the signals that we pass in and out of an instantiated component, except that in a generic's case, the value passed is a constant, and the direction is always "down" or, rather, "in" to the component. We use generics to configure some aspect of a particular instantiation of a component. You might imagine two instantiations of the same component, where one might load a counter with one value, and the other instantiation with another value. In our case, we're telling the component (our target game design) that it's operating in simulation mode. We'll see how this works soon. Note that, like the regularly passed port signals, the generics are identified with the key words "generic map".

　　The other new point is that we declared the "simulation" generic above as a "boolean" type, which can take either a "true" or a "false" value. Here, we assign it as "true", since we want the instantiated component to indeed operate in its simulation mode. We could have declared "simulation" as, say, std_logic, and let a '1' mean simulation mode, and a '0' as non-simulation mode, but this is a bit clearer. We'll see how a boolean signal works when we dive into the operation of the modified target game component design.

　　Since the design we're simulating is the target game that we already covered, we'll simply look at what we've changed in order to facilitate the simulation. We'll look at the entity declaration first:

```
-- -------------------------------------------------
--   Useful information goes here at the top.
-- -------------------------------------------------

library IEEE;
use IEEE.STD_LOGIC_1164.all;
use IEEE.NUMERIC_STD.all;
use IEEE.STD_LOGIC_MISC.all;
use IEEE.STD_LOGIC_UNSIGNED.all;

entity target_game is        1
  port
    (
    clk           : in   std_logic; --50MHz
    pb_in         : in   std_logic; --low-active
    --
    led           : out std_logic
    );
end entity target_game;
```

original target_game

```
-- -------------------------------------------------
--   Useful information goes here at the top.
-- -------------------------------------------------

library IEEE;
use IEEE.STD_LOGIC_1164.all;
use IEEE.NUMERIC_STD.all;
use IEEE.STD_LOGIC_MISC.all;
use IEEE.STD_LOGIC_UNSIGNED.all;

entity target_game_sim is      1
  generic
    ( simulation : boolean := false      2
    );
  port
    (
    clk           : in   std_logic; --50MHz
    pb_in         : in   std_logic; --low-active
    --
    led           : out std_logic
    );
end entity target_game_sim;
```

new target_game_sim

Fig. 4-9

1) We've changed the name of the new simulation-capable design (entity). We didn't really need to (since they're two different files on my computer, located in two different directories), but it avoids confusion;

2) The new version includes the "simulation" generic as we just saw. Note that the simulation mode within the target game module is set to "false". This is an important point. The generic assignment in the file that's doing the instantiating takes precedence—overrides— whatever may be assigned down here. Since "target_game_tb" sets the "simulation" generic to true, that is the value that is used here in "target_game_sim". Why is it set to false here? This serves as more or less a safety default. If we use this file somewhere where it's the top module in the hierarchy, where it's not instantiated by a higher-level module, then we don't necessarily want it in simulation mode.

Simulation changes to submodule declarations

Next we'll examine the differences in the sub-module component declarations.

```
component debounce is
port
  (
    clk             : in  std_logic;
    pb_in           : in  std_logic; --push-button input
    ms_tick         : in  std_logic; --one millisecond indication
    pb_debnced      : out std_logic  --push-button debounced output
  );
end component;

component timer is
port
  (
    clk             : in  std_logic;
    timer_load      : in  std_logic_vector(15 downto 0); --units of milliseconds
    timer_start     : in  std_logic; --begin the timer
    ms_tick         : in  std_logic; --one millisecond indication
    timer_blink     : out std_logic; --one bit of the timer (toggling)
    timer_done      : out std_logic  --indication the the timer has finished
  );
end component;
```

original target_game

```
component debounce is
generic
  ( simulation : boolean
  );
port
  (
    clk             : in  std_logic;
    pb_in           : in  std_logic; --push-button input
    ms_tick         : in  std_logic; --one millisecond indication
    pb_debnced      : out std_logic  --push-button debounced output
  );
end component;

component timer is
generic
  ( simulation : boolean
  );
port
  (
    clk             : in  std_logic;
    timer_load      : in  std_logic_vector(15 downto 0); --units of milliseconds
    timer_start     : in  std_logic; --begin the timer
    ms_tick         : in  std_logic; --one millisecond indication
    timer_blink     : out std_logic; --one bit of the timer (toggling)
    timer_done      : out std_logic  --indication the the timer has finished
  );
end component;
```

new target_game_sim

Fig. 4-10

Here we find the same "simulation" generics as in the "target_gam_tb"/"target_game_sim" pair, since, as we'll see, we want these sub-modules to operate in a special simulation mode as well.

We're expanding the set of constants:

```
constant MS_LOAD        : unsigned(15 downto 0)            := X"C350"; --1ms at 50MHz
constant TAR_TIME       : std_logic_vector(15 downto 0) := X"0180"; --384 milliseconds
constant BLINK_TIME     : std_logic_vector(15 downto 0) := X"0800"; --2K (2 seconds)
constant PAUSE_TIME     : std_logic_vector(15 downto 0) := X"1000"; --4K (4 seconds)
```

original target_game   **1**

2

```
constant MS_LOAD_REAL    : unsigned(15 downto 0)            := X"C350"; --1ms at 50MHz
constant MS_LOAD_SIM     : unsigned(15 downto 0)            := X"0002"; --
constant TAR_TIME_REAL   : std_logic_vector(15 downto 0) := X"0180"; --384 milliseconds
constant TAR_TIME_SIM    : std_logic_vector(15 downto 0) := X"0003"; --
constant BLINK_TIME_REAL : std_logic_vector(15 downto 0) := X"0800"; --2K (2 seconds)
constant BLINK_TIME_SIM  : std_logic_vector(15 downto 0) := X"0006"; --
constant PAUSE_TIME_REAL : std_logic_vector(15 downto 0) := X"1000"; --4K (4 seconds)
constant PAUSE_TIME_SIM  : std_logic_vector(15 downto 0) := X"0002"; --
--
```

```
signal MS_LOAD      : unsigned(15 downto 0);
signal TAR_TIME     : std_logic_vector(15 downto 0);    **3**
signal BLINK_TIME   : std_logic_vector(15 downto 0);
signal PAUSE_TIME   : std_logic_vector(15 downto 0);
```

new target_game_sim

Fig. 4-11

1) Where we originally had four constants,

2) … we now have twice that many. You can see that we have a "real" and a "sim" version of each, where the "sim" versions are far smaller. We'll understand why when we get into the simulation itself.

3) Since we will need to select between the "real" and "sim" versions (using the "simulation" generic, as we'll see), we need intermediate signals to assign one or the other to.

And here in the beginning of the architecture body we make that selection:

```
begin

sim_select : process(clk)
begin
   if (simulation) then
      MS_LOAD     <= MS_LOAD_SIM;
      TAR_TIME    <= TAR_TIME_SIM;
      BLINK_TIME  <= BLINK_TIME_SIM;
      PAUSE_TIME  <= PAUSE_TIME_SIM;
   else
      MS_LOAD     <= MS_LOAD_REAL;
      TAR_TIME    <= TAR_TIME_REAL;
      BLINK_TIME  <= BLINK_TIME_REAL;
      PAUSE_TIME  <= PAUSE_TIME_REAL;
   end if;
end process;
```

new target_game_sim

Fig. 4-12

I expect that this is self-explanatory.

The changes in the pseudo-random section will take a little explaining.

```
pseudo_random : process(clk)
begin
   if rising_edge(clk) then
      if (state = create) then
                     ------------ bit 1 ------------    -- bits 2-10 ----
         pseudo_sr <= (pseudo_sr(7) XOR pseudo_sr(10)) & pseudo_sr(1 to 9);
      end if;
   end if;
end process;

pseudo_seg <= pseudo_sr(1 to 4);

span <= X"0200" when pseudo_seg = X"0" else
     "000" & pseudo_seg & '0' & X"00";
```

original target_game

```
pseudo_random : process(clk)
begin
   if rising_edge(clk) then
      if (state = create) then
                     ------------ bit 1 ------------    -- bits 2-10 ----
         pseudo_sr <= (pseudo_sr(7) XOR pseudo_sr(10)) & pseudo_sr(1 to 9);
      end if;
   end if;
end process;

pseudo_seg <= pseudo_sr(1 to 4);

span <= X"0002"                  when (simulation AND pseudo_seg = X"0") else
     (X"000" & pseudo_seg)       when simulation else
     X"0200"                     when pseudo_seg = X"0" else
     ("000" & pseudo_seg & '0' & X"00");
```

new target_game_sim

Fig. 4-13

In the original file, the four bits of pseudo-random values were followed by nine bits of zeros when assigned to "span". This has the effect of multiplying the 4-bit pseudo-random value by 2^9, or 512. Since the units of "span" are milliseconds, this means that the pseudo-random value is from 1/2 second up to eight seconds—something appropriate for humans to view lit LEDs. This is far too long for simulation, however, and so when in simulation mode, we don't multiply the pseudo-random value at all (it lies at the LS position of "span").

And finally for the target_game_sim file, we include the "simulation" generic to the sub-module instantiations.

```
debounce_ins : debounce
port map
  (
    clk             => clk,          --in  std_logic;
    pb_in           => pb_in,        --in  std_logic;
    ms_tick         => ms_tick,      --in  std_logic;
    pb_debnced      => pb_debnced    --out std_logic
  );

timer_inst : timer
port map
  (
    clk             => clk,          --in  std_logic;
    timer_load      => timer_load,   --in  std_logic_vector(15 downto 0);
    timer_start     => timer_start,  --in  std_logic;
    ms_tick         => ms_tick,      --in  std_logic;
    timer_blink     => timer_blink,  --out std_logic;
    timer_done      => timer_done    --out std_logic
  );
```

original target_game

```
debounce_ins : debounce
generic map
  (
    simulation => simulation
  )
port map
  (
    clk             => clk,          --in  std_logic;
    pb_in           => pb_in,        --in  std_logic;
    ms_tick         => ms_tick,      --in  std_logic;
    pb_debnced      => pb_debnced    --out std_logic
  );

timer_inst : timer
generic map
  (
    simulation => simulation
  )
port map
  (
    clk             => clk,          --in  std_logic;
    timer_load      => timer_load,   --in  std_logic_vector(15 downto 0);
    timer_start     => timer_start,  --in  std_logic;
    ms_tick         => ms_tick,      --in  std_logic;
    timer_blink     => timer_blink,  --out std_logic;
    timer_done      => timer_done    --out std_logic
  );
```

new target_game_sim

Fig. 4-14

Note that we have not explicitly declared "simulation" to be a constant, but this is implied since we carried it down in the target_game_sim file's generic.

Simulation changes to the timer submodule

Next we'll look at the changes made to the two sub-modules for simulation. The first is the timer:

```vhdl
-- ------------------------------------------------------------
--   Useful information goes here at the top.
-- ------------------------------------------------------------

library IEEE;
use IEEE.STD_LOGIC_1164.all;
use IEEE.NUMERIC_STD.all;
use IEEE.STD_LOGIC_MISC.all;
use IEEE.STD_LOGIC_UNSIGNED.all;

entity timer is
  generic
    ( simulation : boolean := false          1
    );
  port
    (
      clk            : in  std_logic;
      timer_load     : in  std_logic_vector(15 downto 0); --ms units
      timer_start    : in  std_logic;
      ms_tick        : in  std_logic;
      timer_blink    : out std_logic;
      timer_done     : out std_logic
    );
end entity timer;

architecture Behavioral of timer is

    signal timer_count    : unsigned(15 downto 0) := X"0000";

begin

    timer_process : process(clk)
    begin
      if rising_edge(clk) then
        if (timer_start = '1') then
          timer_count <= unsigned(timer_load);
        elsif (ms_tick = '1') then
          if (timer_count /= "0000") then
            timer_count <= timer_count - 1;
          end if;
        end if;
        --
        if (    ms_tick = '1'
            and timer_count = "0001"
           ) then
          timer_done <= '1';
        else
          timer_done <= '0';
        end if;
      end if;
    end process;
                                              2
    timer_blink <= timer_count(0)  when simulation else
                   timer_count(8); --blinks every 1/4 second

end architecture Behavioral;
```

new timer

Fig. 4-15

1) we bring in the generic "simulation" in the same manner as the target_game_sim file;

2) we modify the "timer_blink" output.

Here's the before-and-after:

1

```
timer_blink <= timer_count(8); --blinks every 1/4 second
```

original target_game

2

```
timer_blink <= timer_count(0)  when simulation else
               timer_count(8);  --blinks every 1/4 second
```

new target_game_sim

Fig. 4-16

1) The original design used bit 8 of the timer counter for the blinking LED (a successful target "hit"). Since the timer decrements every millisecond, bit 8 blinks (toggles on and off) every 2^8 milliseconds, or 1/4 second.

2) A quarter second is essentially forever in a simulation, so, for simulations, we blink on and off each "millisecond" (in quotes, since even the millisecond duration is vastly reduced).

The Modelsim simulation tool

We now have a testbench, and a design that's been modified to accommodate simulation, and we're ready to actually do a simulation. I am using the Modelsim simulation software, but principles should carry to other types.

This is the Modelsim window upon first launching:

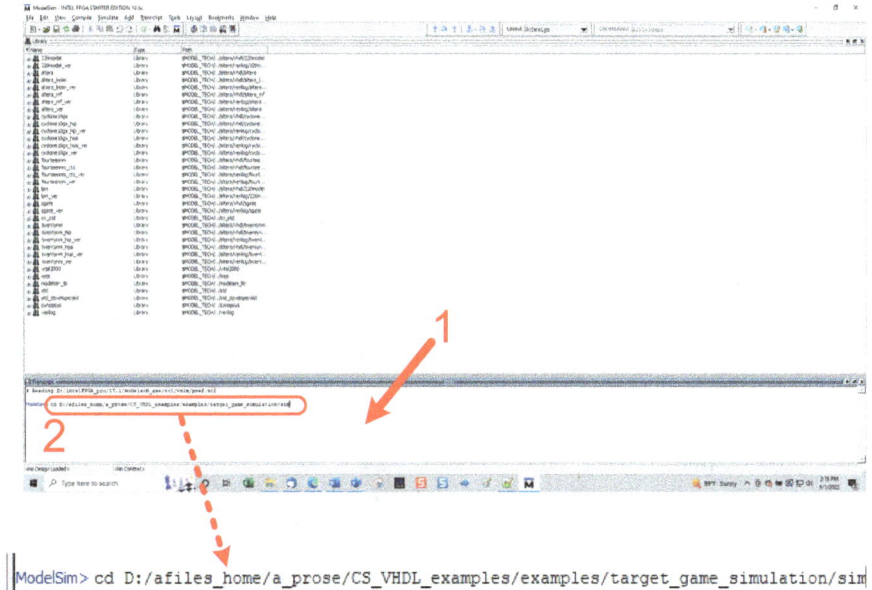

Fig. 4-17

1) This is the transaction window, where you apply your commands (we'll be using the command-based mode, versus strictly GUI);

2) The first thing you'll want to do is move to your simulation folder—this is the standard "cd C://…" form.

But before continuing, we should look at my folder structures:

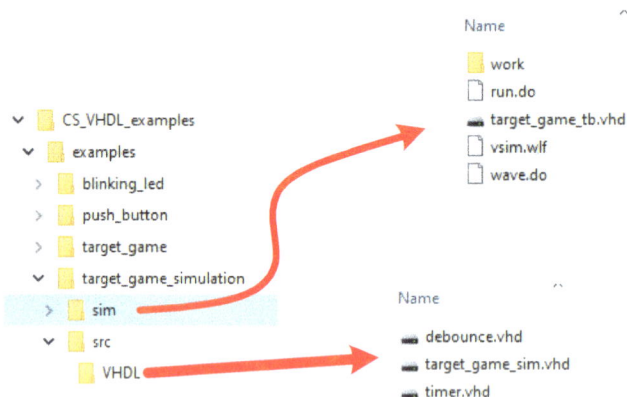

Fig. 4-18

Underneath the "target_game_simulation" main example project folder, I created two sub-folders, one called "sim", and the other "src/VHDL". The reason for the "src" folder is that sometimes we have types of source files other than VHDL, for example IP modules provided by other vendors. Within the "VHDL" folder are the three VHDL comprising this design.

The "sim" folder is dedicated to simulation, and these are the components:

"work" folder: created and used by Modelsim. You may have to use the GUI and Compile just the testbench file first to get Modelsim to create it;

"run.do": I've created this, and it's basically a runtime batch file—we'll get to it shortly;

"target_game_tb.vhd": our testbench;

"vsim.wlf": created and used by Modelsim;

"wave.do": created and used by Modelsim, but we have direct input, as we'll see.

Other folder structures are possible; this is just the one I use.

Simulation run.do file

Next we'll look at the contents of the "run.do" file. Note that the file name can be pretty much anything, I chose this one since it makes sense (to me).

```
vcom -work work -2002 -explicit ../src/VHDL/timer.vhd
vcom -work work -2002 -explicit ../src/VHDL/debounce.vhd
vcom -work work -2002 -explicit ../src/VHDL/target_game_sim.vhd
vcom -work work -2002 -explicit target_game_tb.vhd

vsim work.target_game_tb
#do wave.do
#run 1.46 us
```

1

2

3

4

Fig. 4-19

1) The "vcom" lines instruct Modelsim to compile each file. You can see in the path names why we looked at my folder structure. During the compile of each file, Modelsim may find problems, and, if so, will explain in the transaction window;

2) "vsim" tells Modelsim to simulate (we've only compiled until now). Here's where Modelsim may find problems in the connections between your modules, and, again, if so, will explain in the transaction window;

3) "do wave.do" tells Modelsim to launch the waveform viewer. This isn't valid until we create a waveform, and so it is commented ("#");

4) The "run" command is obvious, and is also commented for now, since we'll want to create a waveform first.

The next step is to run the batch file by typing "do run.do" in the transaction window. Modelsim will compile the files, and, if successful, will launch the waveform viewer (usually in it's own separate window), but the waveform viewer will be empty initially.

Modelsim "instance" and "objects" windows

Here's the resulting Modelsim windows (assuming there were no errors in your code):

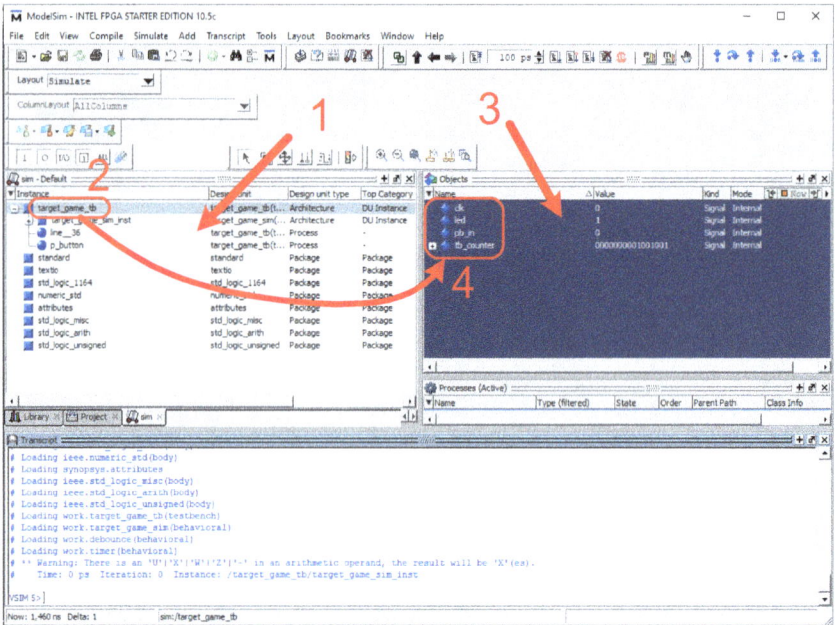

Fig. 4-20

1) The "Instance" window shows the various code components that Modelsim found, but,

2) ... the only important one is your top simulation file, the testbench;

3) This window is the "Objects" window, and for us, where we find the code signals;

4) And, we can see that Modelsim is showing the signals within our testbench. These are important, since this is where we select the signals that we'd like to copy into the waveform window for simulation viewing.

Adding signals to Modelsim waveform windows

And now we'll do that.

This is the window after the first compiling—empty:

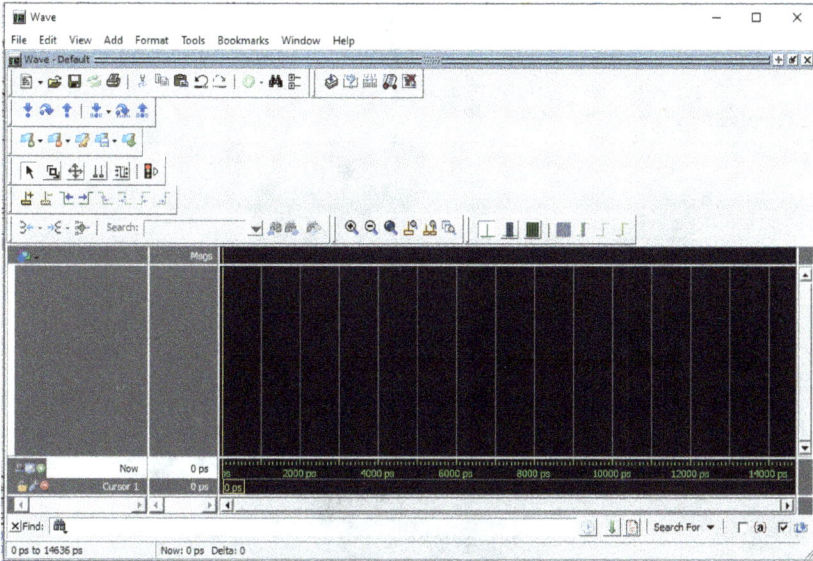

Fig. 4-21

We select and copy all the signal in the testbench (you can select as many or as few as you like):

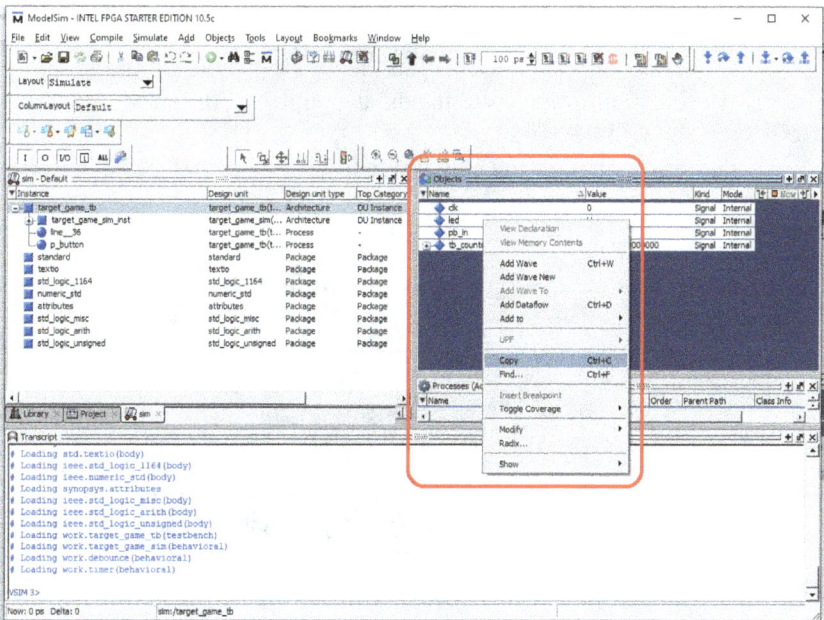

Fig. 4-22

And then paste them into this window of the waveform viewer:

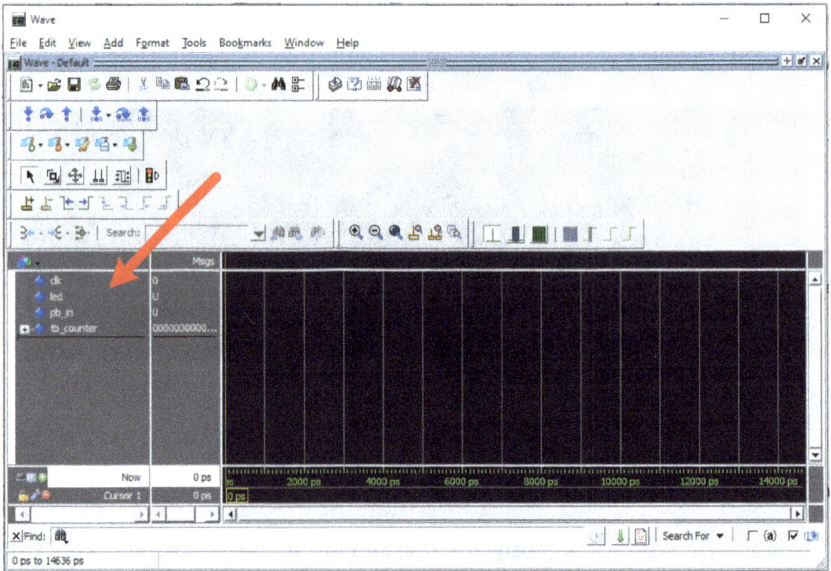

Fig. 4-23

You may need to click within the window before pasting.

Before continuing, we'll add a couple of dividers, which as we'll soon see will be handy.

We right-click in the signal window …

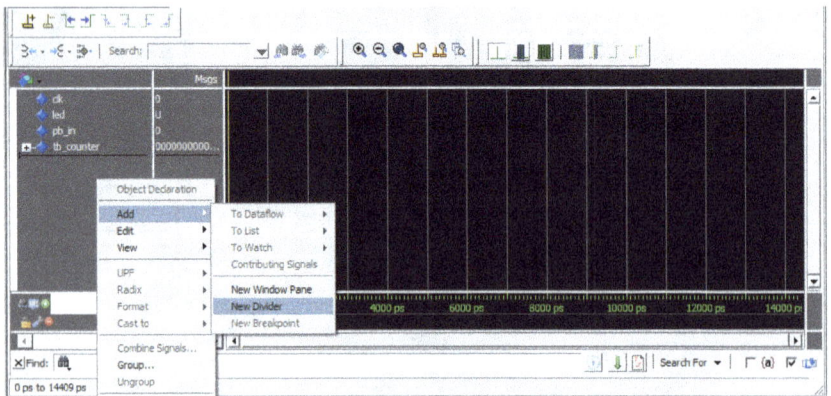

Fig. 4-24

... and select "Add", and "New Divider" ...

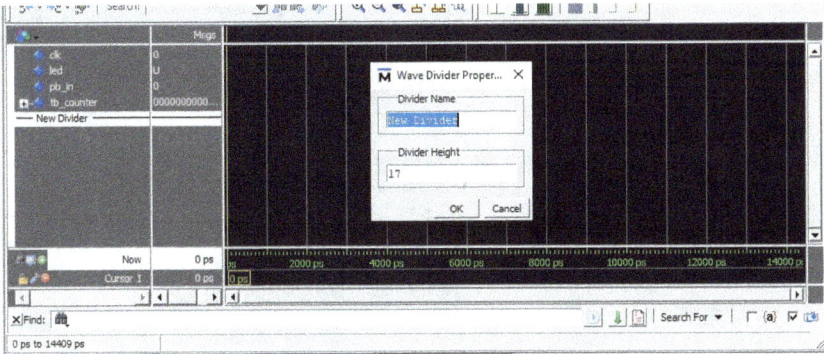

Fig. 4-25

... with this result.

I've entered "target_game_sim" ...

Fig. 4-26

... and then repeated this for three more dividers ("Debounce", "Timer", and "End").

We'll next select some useful signals from the top design module instantiation (target_game_sim) ...

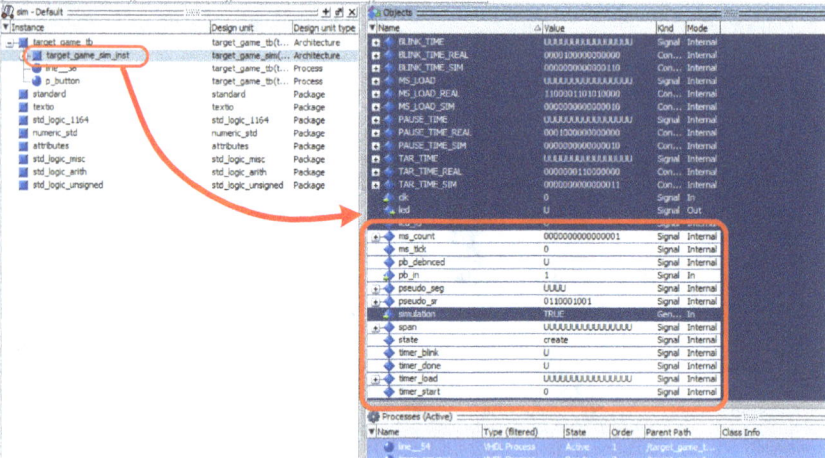

Fig. 4-27

… and copy them to the waveform:

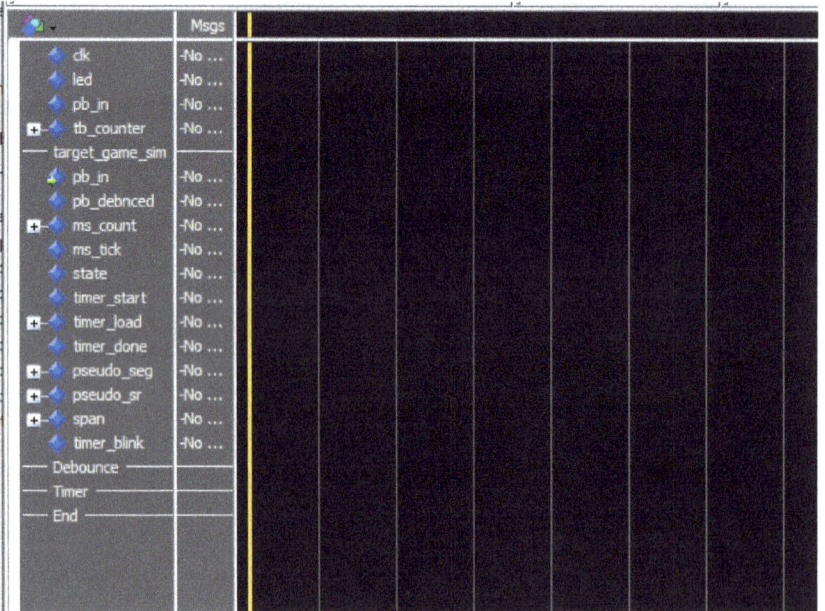

Fig. 4-28

Notice that Modelsim inserts the new signals—the debounce signals—*above* where you select. Thus, having the dividers in place works well to keep the signals positioned where you like when you copy them in. Note that I've also rearranged their order to follow the flow of the logic operation (by simply selecting them and moving up or down).

We'll next expand the top design module instantiation (target_game_sim):

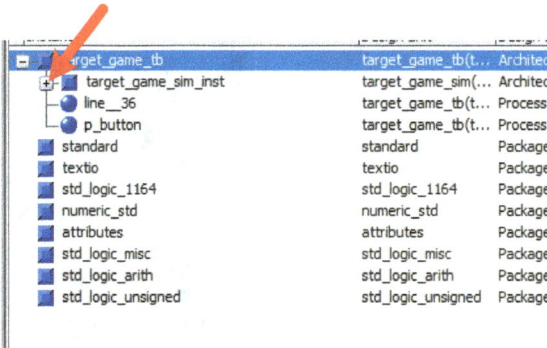

Fig. 4-29

With this result:

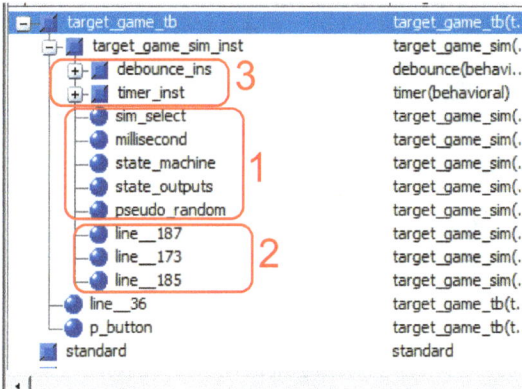

Fig. 4-30

1) These are the various processes in the code—we don't really care about these;

2) And, these are the non-process assignments, and we don't care about these either;

3) These are the instantiated sub-modules, and these are indeed of interest.

From here, we can expand each instantiated sub-module ("debounce" here, for example), and select some signals to view …

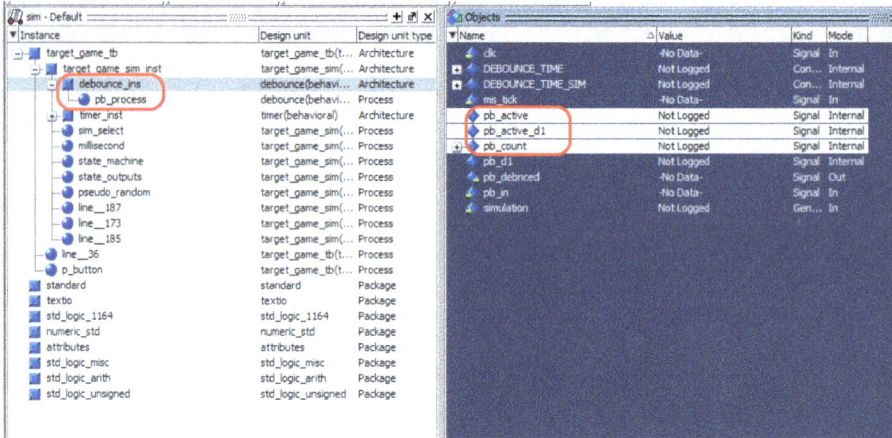

Fig. 4-31

… starting with the "debounce" module, and copy them to the waveform.

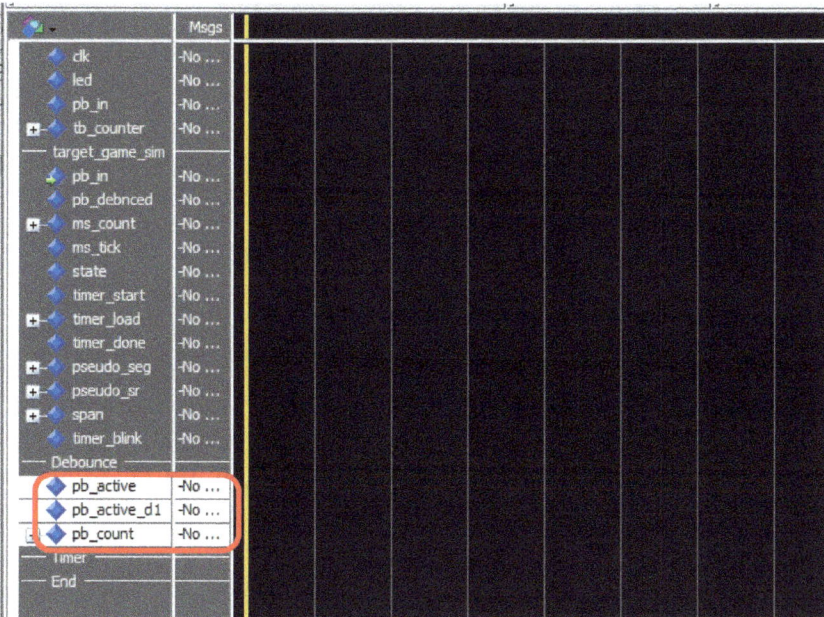

Fig. 4-32

Next, I copy a useful signal from the Timer module …

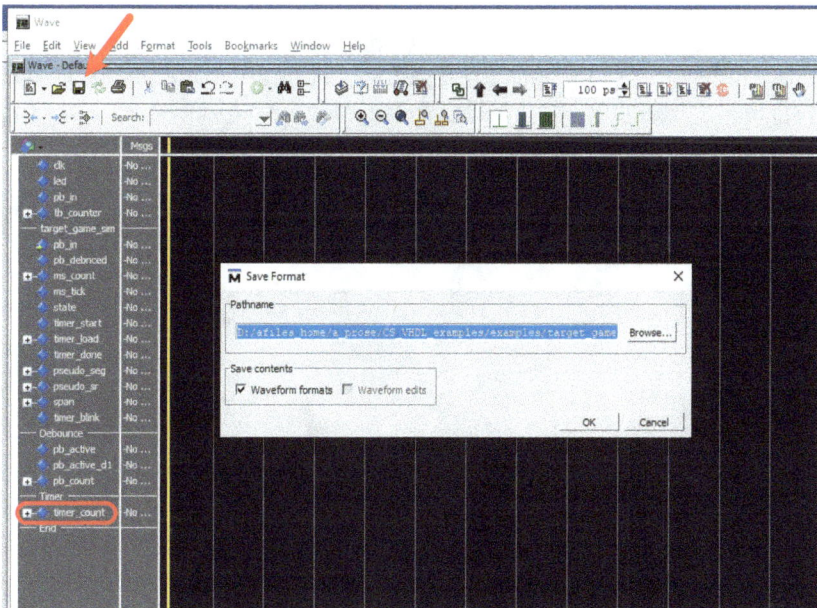

Fig. 4-33

… and am then doing an important step—saving the changes. This becomes your "wave.do" file that Modelsim will create when you save. Modelsim defaults to its project working folder with the name "wave" (as we've already seen), but you can redirect where it's stored if you like by changing the path name that appears and/or the file name (something other than "wave").

Running a Modelsim simulation

Since I am letting Modelsim save the wave file in the sim project working folder, my command to open the waveform using the wave file in my batch file includes no path name.

```
vcom -work work -2002 -explicit ../src/VHDL/timer.vhd
vcom -work work -2002 -explicit ../src/VHDL/debounce.vhd
vcom -work work -2002 -explicit ../src/VHDL/target_game_sim.vhd
vcom -work work -2002 -explicit target_game_tb.vhd

vsim work.target_game_tb
do wave.do
run 1.46 us
```

Fig. 4-34

We uncomment the "do wave.do " and "run 1.46 us". This last tells Modelsim to simulate for that long, noting that the clock in our testbench is running at 50 MHz, so we're telling Modelsim to run for 73 clock periods.

Entering "do run.do" in the transaction window, we get:

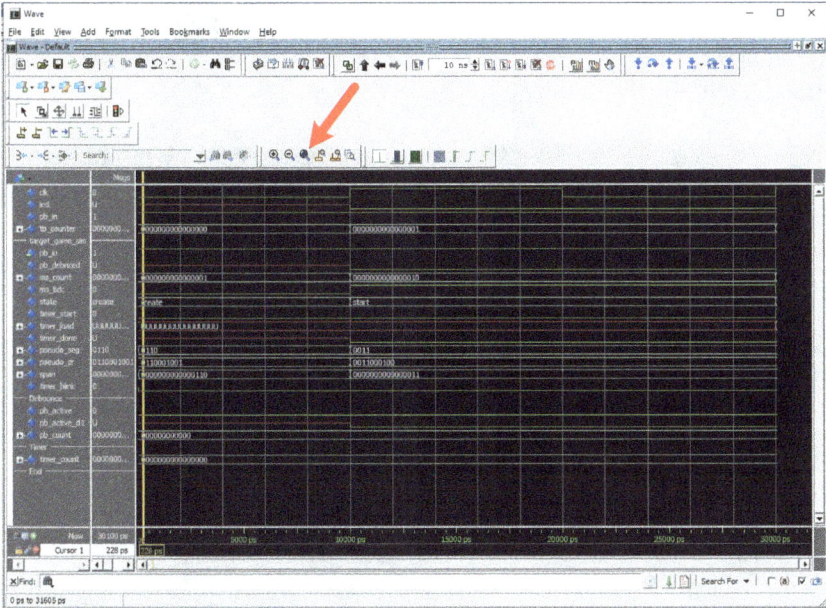

Fig. 4-35

… which doesn't look like much, but that's only because we're looking at a tiny portion of the full waveform. But the solid magnifying glass selects the entire span, and if we click on it, we get:

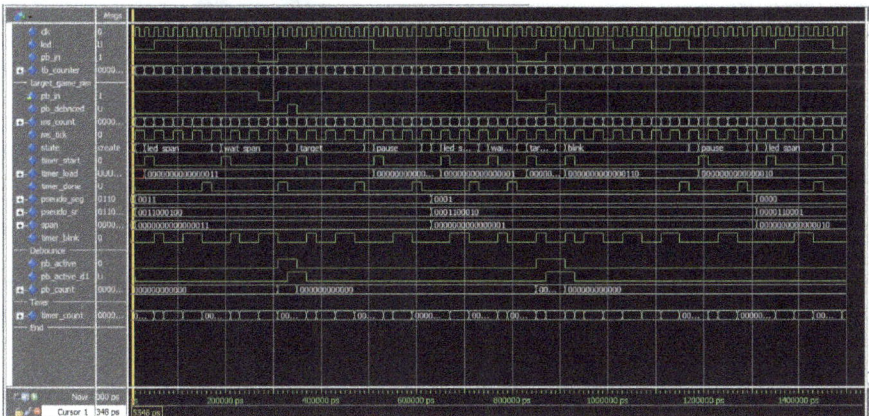

Fig. 4-36

There's one more tweak we'll make before looking at the simulation results. You'll notice that the displayed values default to binary, e.g., pb_count.

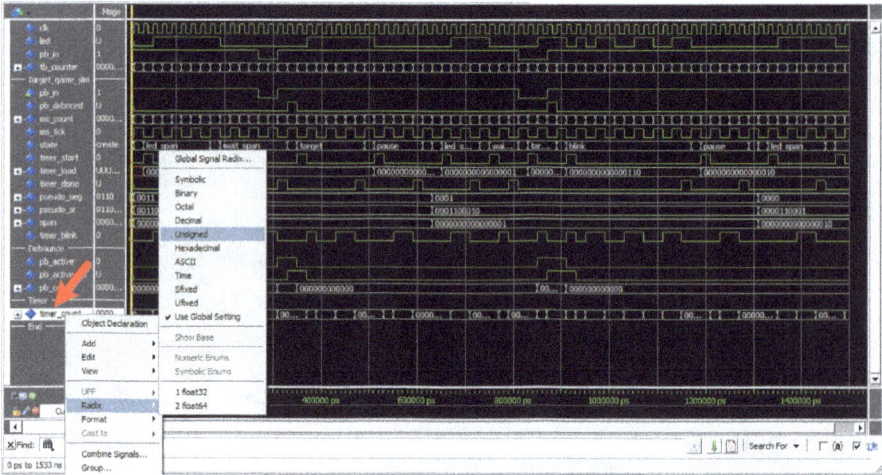

Fig. 4-37

If I right-click on, for example, "pb_count", and select Radix/Unsigned …

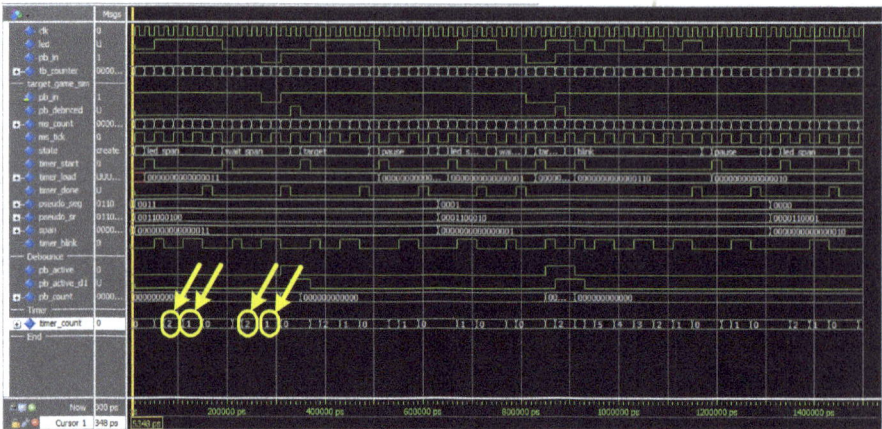

Fig. 4-38

… the values in the waveform window are now displayed as decimal (actually unsigned, but this appears as decimal).

Examining the simulation results

We're finally ready to examine the simulation results. We begin by zooming into the early time:

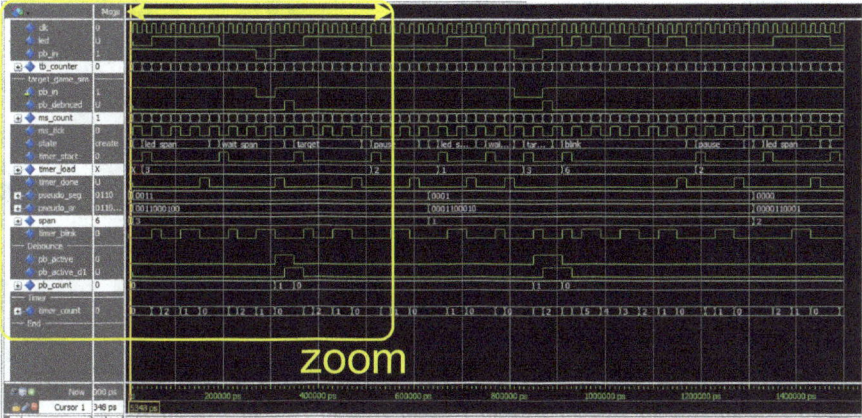

Fig. 4-39

Note that I've converted all the count related signals to unsigned. Here's the zoomed image:

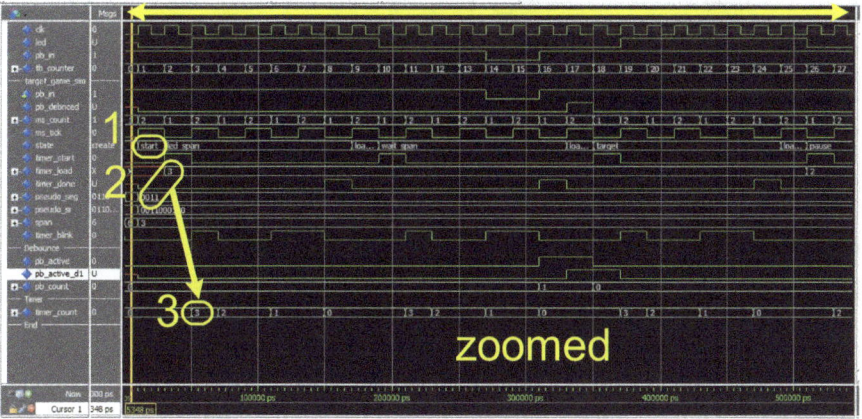

Fig. 4-40

We can see that the initial value of "pseudo_seg" is 3 ("0011").
1) The initial "start" state,
2) loads the "pseudo_seg" 3 value …
3) into "timer_count" in the timer module.

And …

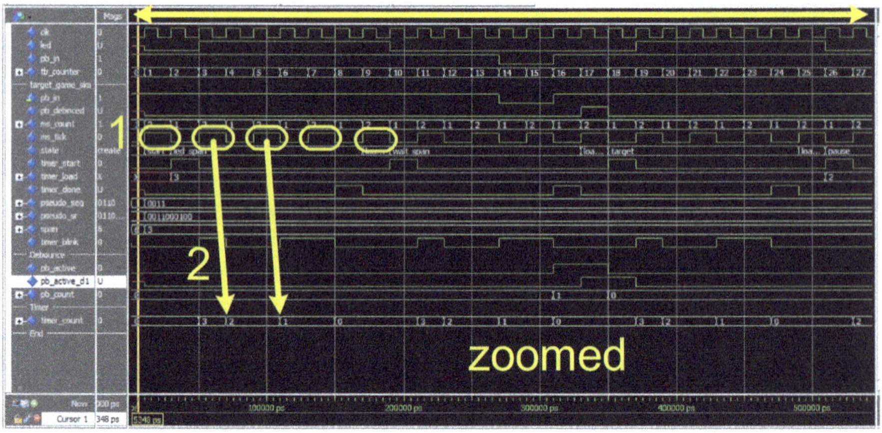

Fig. 4-41

1) We've configured the millisecond ticks ("ms_tick") to occur every other clock, i.e., every 40 nanoseconds, instead of the real-time one thousand nanoseconds,
2) and thus, "timer_count" decrements every other clock as well.

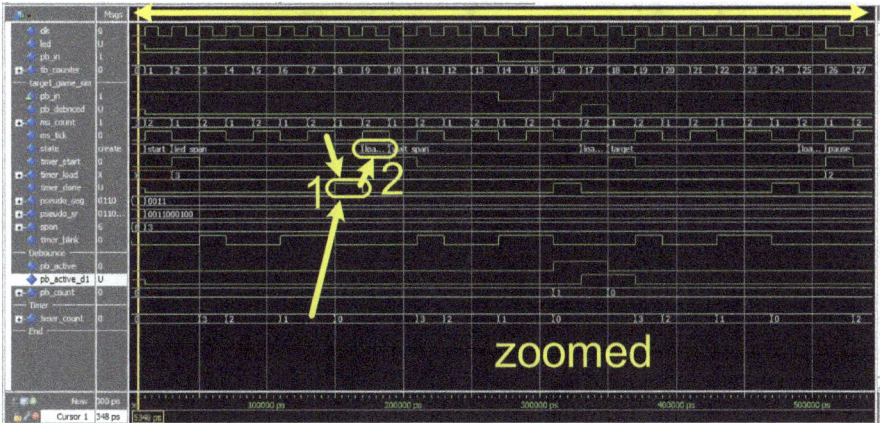

Fig. 4-42

1) The combination of "ms_tick" and a "timer_count" of one activates the timer module's "timer_done", and

2) moves the state machine to the "load_wait" state …

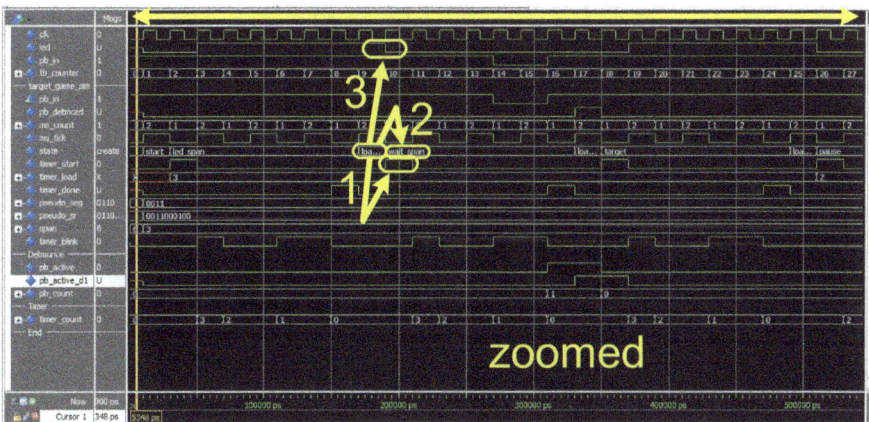

Fig. 4-43

1) … which then activates "timer_start",

2) the sate machine steps ahead to the "wait_span" state, and

3) turns off the LED.

In a similar fashion,

97

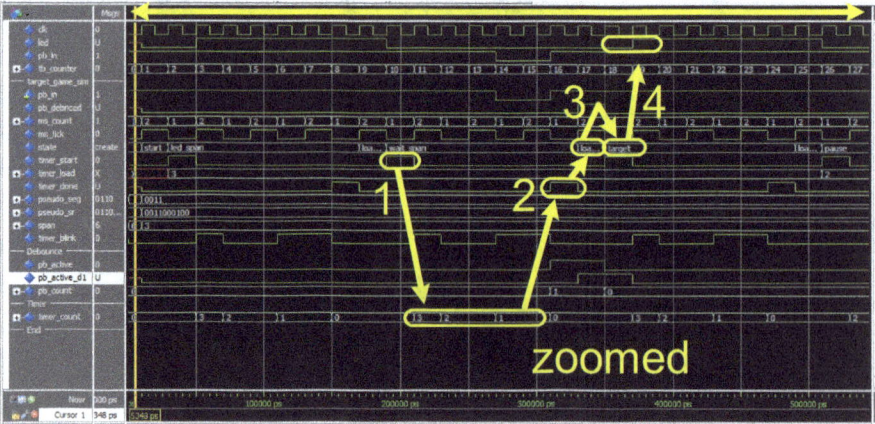

Fig. 4-44

1) the "load_wait" state activates "timer_start", which loads the "timer_count",

2) which activates "timer_done" when finished,

3) which steps the state machine to the "load_target" state, which

4) then activates the LED.

Now that we have a general sense of how the timer and state machine work together, we'll zoom out to the full waveform, and look at the overall operation of the game:

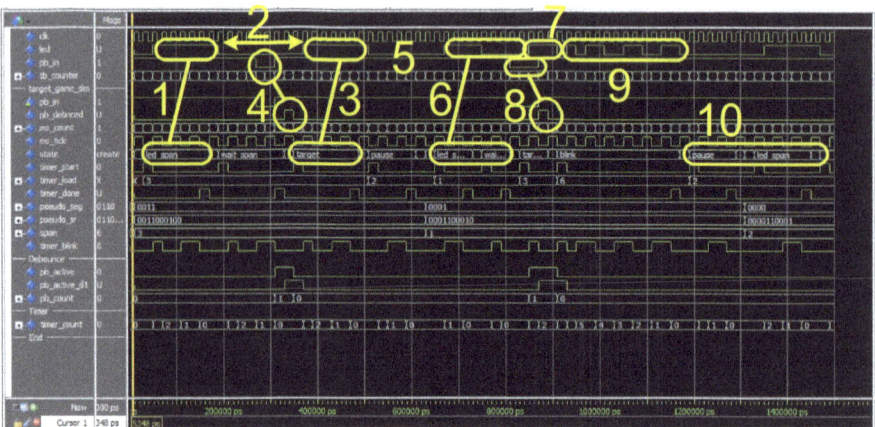

Fig. 4-45

1) We've seen that the LED is initially lit for a (somewhat) randomly determined time,

2) followed by what is supposed to be an equal "wait" time. If you count the clocks, however, you see that the LED is lit for seven clocks, but the following "wait" time (unlit) is nine clocks. The difference is due to the fact that the timer count is enabled by the millisecond tick (ms_tick), which, when interacting with the state machine, produces some imprecision in the entry and exit of the spans, some fuzziness, if you will. Keep in mind that we've drastically condensed the timing of the operation for the sake of simulation (the ms_tick is one clock apart, versus 50,000 for real operation). The fuzziness is a tradeoff for the convenience of simulation. In the actual operation of the game, the fuzziness might be, worst case, two full milliseconds, and for a one-second span, would represent a 0.2% imprecision, hardly something anyone would notice;

3) Per the operation of the game, the "wait" span is then followed by the target span, which the user attempts to "hit" with the push-button ...

4) but here, for the first one, misses (too early);

5) After a pause time,

6) we get the next LED and wait spans,

7) followed by the next target, but the target span is cut short, because ...

8) now the user has hit the target with the push-button, and

9) the game issues a blinking period to demonstrate a successful hit,

10) before proceeding on to the next pause and LED span.

A look at the simulated debounce operation

Before leaving simulations, we'll take a closer look at the push-button operation.

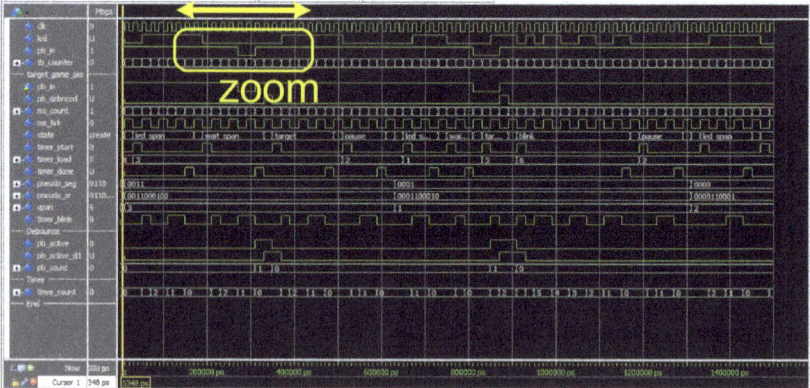

Fig. 4-46

… and we'll zoom in here.

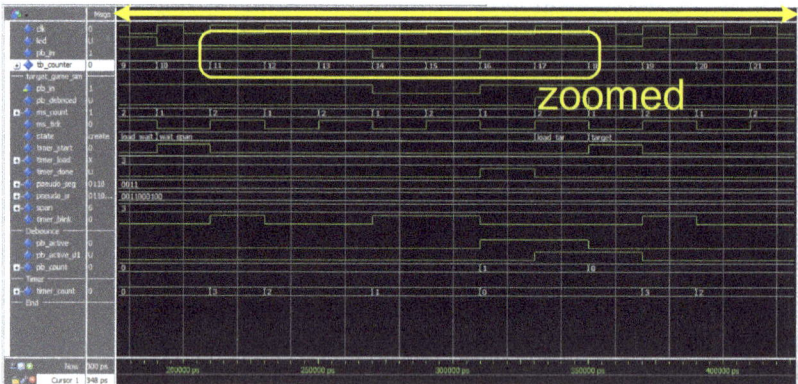

Fig. 4-47

Recall that the testbench sets the pb_in signal when the tb_counter is hex "D" (low-active, remember), which is 13 decimal. And, since pb_in is registered, it appears out of the module at tb_counter 14.

Here's the testbench code:

```
p_button : process(clk)
begin
   if rising_edge(clk) then
      tb_counter <= tb_counter + 1;
      --
      case(tb_counter) is
         when X"000D" => pb_in <= '0';
         when X"000F" => pb_in <= '1';
         when X"0028" => pb_in <= '0';
         when X"002B" => pb_in <= '1';
         when others => null;
      end case;
   end if;
end process;
```

Fig. 4-48

Blaine C. Readler

Chapter 5

UARTs

UARTs have been around for over fifty years, and were once the de facto universal communications method. In fact, the acronym stands for "Universal Asynchronous Receiver, Transceiver." The key here is the word "asynchronous", meaning that the transmission carries along no associated clock, whether a dedicated channel or embedded in the data transitions. Instead, each end is configured to expect the data units to occur at the same clock rate—virtual clock periods—called the BAUD rate. Knowing the virtual clock rate, once the receiving end detects a first transition of data, it can extrapolate the approximate locations of subsequent bit boundaries, and can sample the received stream of data bits at what it supposes is the mid-bit regions. Thus:

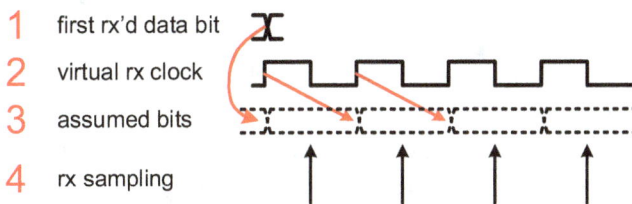

Fig. 5-1

1) Once a first transmission transition is detected,

2) the receiving side assumes a virtual clock proceeding from there,

3) and assumes that the receiving data will have boundaries associated with the virtual clock.

4) The receiver then samples the arriving data at what it assumed to be the data midpoints.

Early on, when UARTs were conceived, the technologies for accurate clock synthesis were limited, and the receiving side of the link could predict the true transmitting clock rate only so accurately. Thus, the receiving virtual clock could drift with respect to the true transmitting clock, like so:

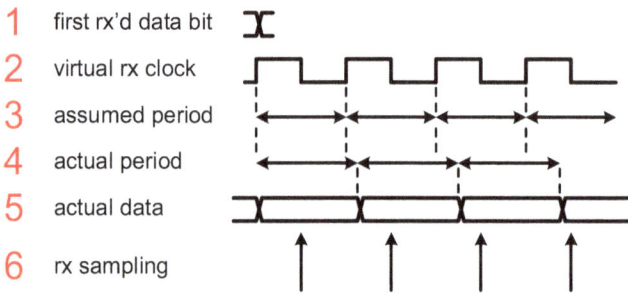

1	first rx'd data bit
2	virtual rx clock
3	assumed period
4	actual period
5	actual data
6	rx sampling

Fig. 5-2

1) a detection of received transition, creates …

2) … an assumed clock, and

3) an associated clock period.

4) The actual clock period …

5) … and associated data, however, might be, for example, longer,

6) so that, while the received sampling works for awhile, it will eventually get out of step and mis-sample the data.

This figure exaggerates the possible inaccuracies between the transmitting and receiving clocks, but because of the possible differences in the transmitting and receiving clocks, each UART transmission is limited to one byte. In other words, each transmitted byte causes the receiving side to recalibrate—a refresh, more or less, of the receiving virtual clock generation.

UART transmissions are serial, meaning that they consist of one wire, carrying one bit at a time. This is the structure:

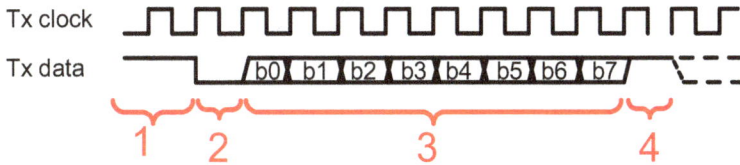

Fig. 5-3

1) During idle times, when there's no data being transmitted, the signal is one, or high;
2) A byte transmission begins with a "start" bit, which is simply a zero, or low,
3) followed by the data, and,
4) finishing with a "stop" bit, which is just a one, or high. The stop bit simply guarantees that the line will be high for enough time for a next "start" bit to be recognized. The "stop" bit is essentially a guaranteed minimum idle time.

We've been assuming a regular 8-bit byte, but UARTs offer variations:

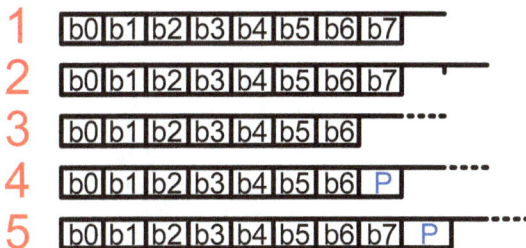

Fig. 5-4

1) An 8-bit byte, with one stop bit;
2) An 8-bit byte, with two stop bits;
3) A 7-bit "byte", with one or two stop bits;

4) A 7-bit "byte", with a parity bit (even or odd), and one or two stop bits;

5) An 8-bit byte, with a parity bit (even or odd), and one or two stop bits.

These variations are typically communicated to the UART device via the following parameters:

o "byte" length (7 or 8 bits);

o number of stop bits;

o parity included;

o parity is even (vs. odd).

Additionally, the assumed clock rate—the BAUD rate (bits-per-second)—must be configured at both ends. There are many values, and here are a few popular ones:

4800

9600

19200

57600

115200

These are just a sampling, and of these, 9600 BAUD is by far the most popular, and is indeed often the default rate.

Two stop bits are offered for receivers that might require additional time to recover and recalibrate, or need time to store away the received data value. Two stop bits is really just a legacy offering, since any devices manufactured in the last few decades will easily operate with just one.

Then there's the issue of 7-bit "bytes". It would be better to talk about "data units" instead of bytes for UARTs. Although it is possible to communicate actual bytes of data (e.g., hexadecimal values), that's only possible if the UART is configured for 8 bits. The majority of data communicated via UARTs is ASCII characters. There are 128 possible combinations in a 7-bit value, and in ASCII format, each combination represents one character, where the characters include essentially all the keys on your keyboard (upper and lower-case letters, numbers, special character like # and $, etc.), but also a whole slew of special characters that were relevant back in the days of teletype machines. Of these, two are still frequently found: Line Feed (LF) and Carriage Return (CR), both of which are needed when displaying text on a screen. ASCII tables are easily found on the internet.

Before we move on to a UART implementation, we should take a moment to understand that UARTs are logical devices—they essentially work with binary states. After all, idle, start, and stop bits are just logical zeros and ones. The actual transmission, once the signal has left the UART device (FPGA or dedicated component), is left to other means. The RS-232 standard is the most common (with RS-485 coming up a distant second), and defines both connector specifics as well as physical signal characteristics. The RS-232 interface specification includes a number of signals associated with modem/computer interactions, for example RTS (Request To Send) and CTS (Clear To Send), but we'll be looking at the simplest (and most common) configuration that includes just one output signal (Tx), and one input signal (Rx).

UART, Tx

We'll start with the UART transmission, since of the two (Tx and Rx), this is the simplest to implement, since on the Tx side we control the operation, whereas on the Rx side, we must adapt to what the other side (their Tx) sends. Further, we'll assume the simplest transmission form: seven bits, no parity, and one stop bit.

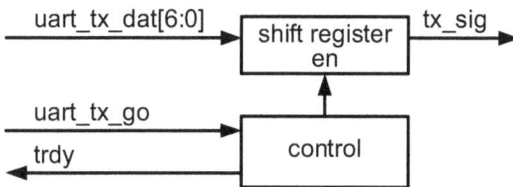

Fig. 5-5

A transmission begins when the user presents a "unit" of data, uart_tx_dat (in this case, a 7-bit ASCII character), and activates uart_tx_go, which indicates that the transmit data is valid and ready to go. The control logic activates the enable of the shift register, which serializes the seven bits, and outputs them one at a time (unusually for most serial transmissions, LS bit first). Once the control logic has determined that all seven bits have been sent, it de-activates the shift register enable, and raises the trdy signal ("transmit ready") to notify the user that a next ASCII character can be sent if desired.

We'll dive into the control logic, but first we need to add one more function:

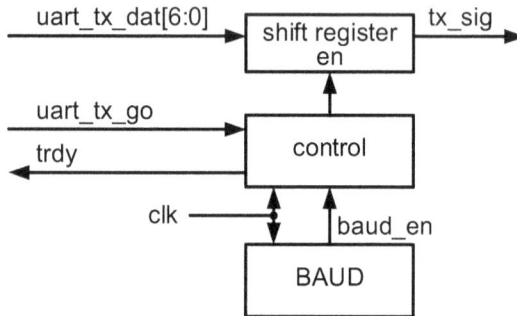

Fig. 5-6

Since our UART interface may need to operate at a configurable BAUD rate, we add some logic to manage this. Both the control and BAUD logic operate off of a relatively fast clock signal (clk), and the baud_en output from the BAUD logic is a one clock-wide enable signal that defines the BAUD rate (baud_en pulses once each BAUD rate cycle).

Here's the transmit half of the UART:

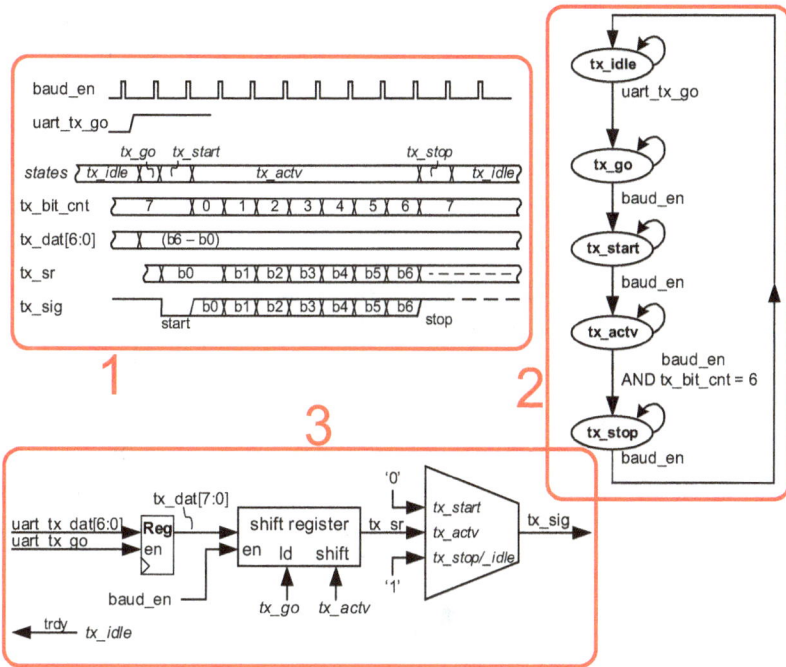

Fig. 5-7

1) A timing diagram, showing one "byte" transmission,
2) the state machine comprising the control function, and
3) a block diagram of the shift register operation.

The core of our design operates at the configured BAUD rate, essentially waking up for one clock period at each baud_en pulse. This allows the same design to accommodate a variety of configurable BAUD rates. However, the signal from the user that initiates a transmission, uart_tx_go, can happen at any time. Thus, when a uart_tx_go occurs, we need to do whatever is necessary to prepare for the next occurring uart_tx_go. In the case of our design, this consists of latching the data to be transmitted.

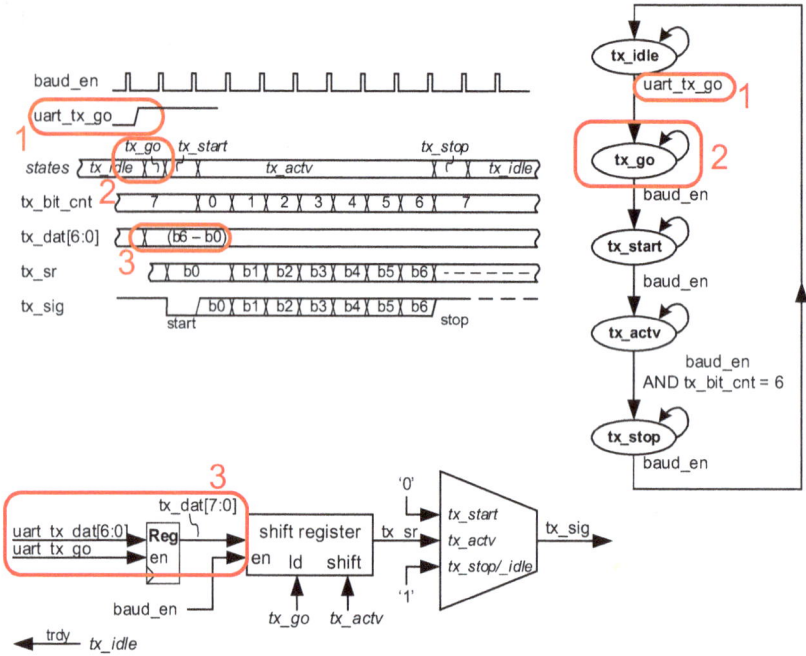

Fig. 5-8

1) A uart_tx_go input, regardless of baud_en, kicks off the state machine,

2) which advances to the tx_go state,

3) and latches the pending 7-bit value to be sent as tx_dat.

The baud_en signal then enables basic operation:

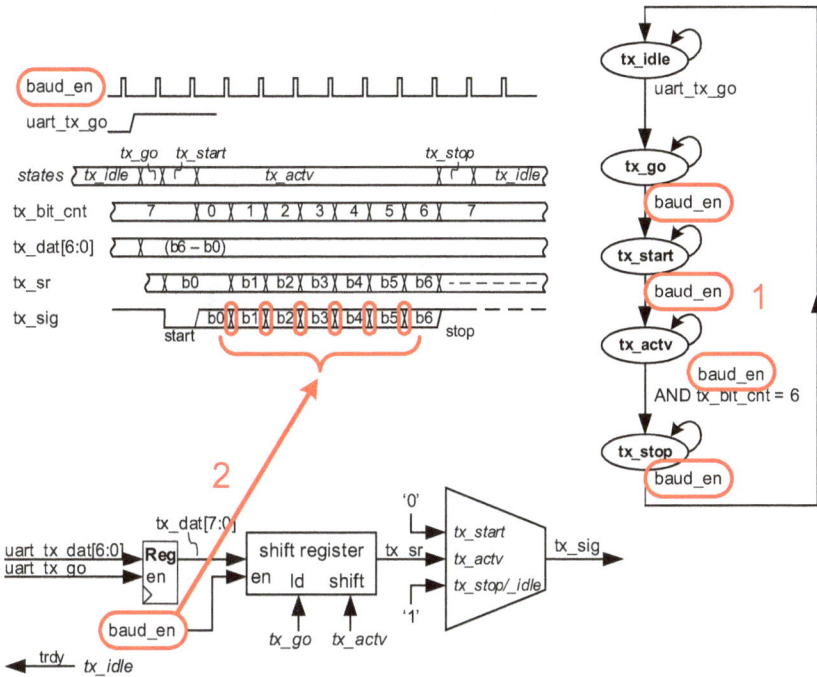

Fig. 5-9

1) including the subsequent steps of the state machine,
2) and the serial shifting of the data bits out of the shift register.

The state machine directly forces the serial output signal low for the start bit, and high for the stop bit:

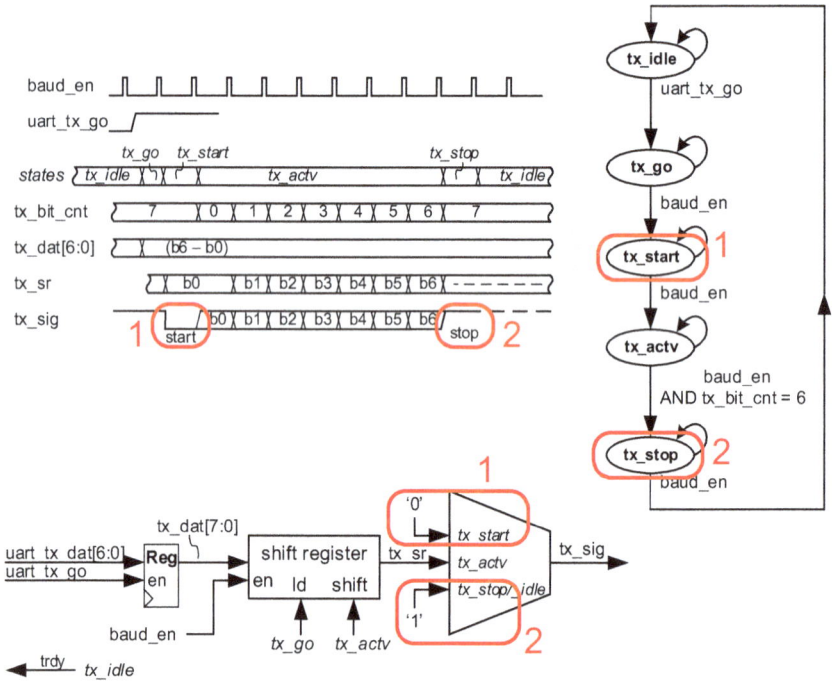

Fig. 5-10

1) the start bit;
2) and the stop bit.

Note that the serial output is also forced to one during the state machine's idle state.

A counter, tx_bit_cnt, tracks the serial bits as they're shifted out:

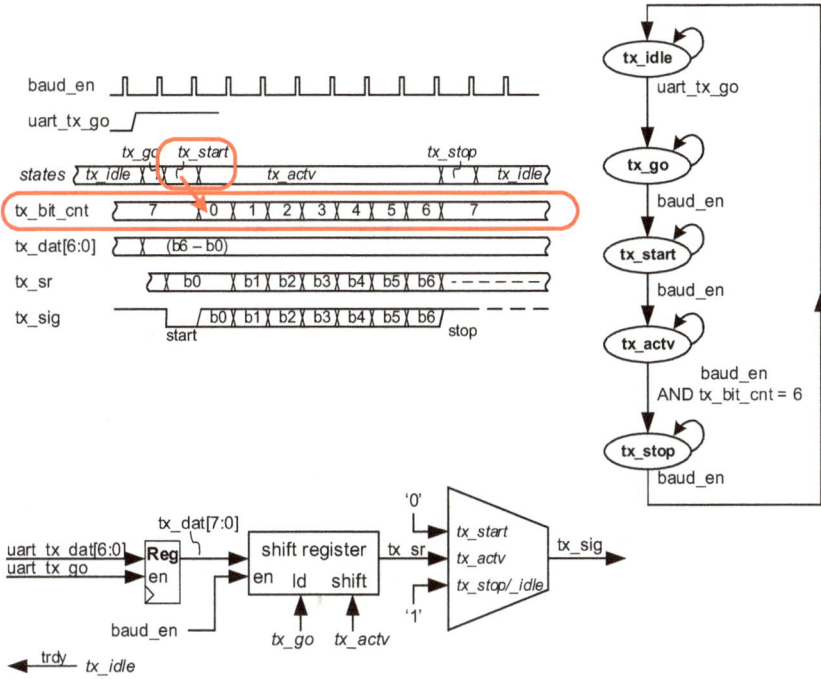

Fig. 5-11

The counter is cleared by the state machine's tx_start state:

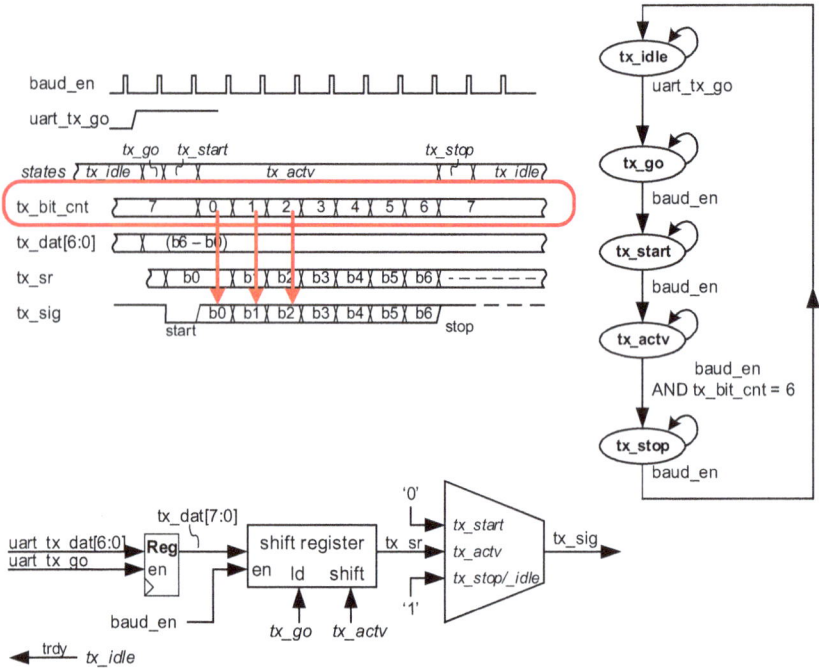

Fig. 5-12

and then tracks each serial bit …

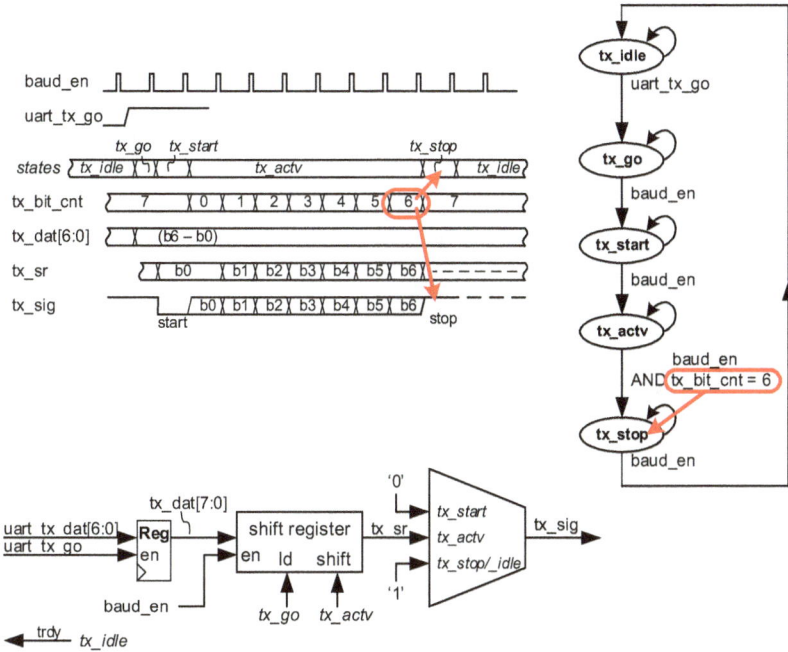

Fig. 5-13

... until the last bit allows the state machine to step ahead to the stop bit.

UART, Tx code

Next we'll look at the VHDL code that implements the Tx half of the UART, and we start with the entity and architecture declarations:

```
library IEEE;
use IEEE.STD_LOGIC_1164.all;
use IEEE.NUMERIC_STD.all;
use IEEE.STD_LOGIC_MISC.all;
use IEEE.STD_LOGIC_UNSIGNED.all;

entity uart_tx is                                                      1
  port (
        clk             : in    std_logic;
        uart_tx_dat     : in    std_logic_vector(6 downto 0);
        uart_tx_go      : in    std_logic;
        tx_sig          : out   std_logic;
        trdy            : out   std_logic := '0'
        );
end entity uart_tx;

architecture Behavioral of uart_tx is
                                                                       2
    -- baud constants assume a 50MHz clock
    constant BAUD_9600 : std_logic_vector(15 downto 0) := X"1458";

    component baud is
      port (                                                           3
            clk             : in    std_logic;
            baud_config     : in    std_logic_vector(15 downto 0);
            baud_en         : out   std_logic
          );
    end component;

    type tx_sm_type is (
                          tx_idle,
                          tx_go,
                          tx_start,                4
                          tx_actv,
                          tx_stop
                        );
    signal tx_sm : tx_sm_type;

    signal baud_en          : std_logic;
    signal tx_bit_cnt       : unsigned(3 downto 0) := X"0";
    signal tx_dat           : std_logic_vector(6 downto 0) := "0000000";
    signal tx_sr            : std_logic_vector(6 downto 0) := "0000000";

begin
```

Fig. 5-14

1) The entity I/O signals, as we saw in the block diagram;

2) The constant that establishes our 9600 BAUD rate. Note that X"1458" is decimal 5,208, which, divided into the 50MHz clock rate, equals 9,600.6;

3) The component that creates the BAUD enable signal (baud_en); and

4) the states of the state machine from the previous diagram.

We'll split the architecture body into two parts. The first half:

```vhdl
begin

   baud_inst : baud                                                    1
     port map
       (
         clk            => clk,        --in  std_logic;
         baud_config    => BAUD_9600,  --in  std_logic_vector(15 downto 0);
         baud_en        => baud_en     --out std_logic
       );

   uart_tx_bit_cnt : process(clk)
   begin                                          2
     if rising_edge(clk) then
       if (baud_en = '1') then
         if (tx_sm = tx_start) then
           tx_bit_cnt <= X"0";
         elsif (tx_bit_cnt /= X"7") then
           tx_bit_cnt <= tx_bit_cnt + 1;
         end if;
       end if;
     end if;
   end process;

   uart_tx_sm : process(clk)                                               3
   begin
     if rising_edge(clk) then
       case (tx_sm) is
         when tx_idle  => if (uart_tx_go = '1') then     tx_sm <= tx_go;      end if;
         when tx_go    =>   if (baud_en = '1') then      tx_sm <= tx_start;   end if;
         when tx_start => if (baud_en = '1') then        tx_sm <= tx_actv;    end if;
         when tx_actv  => if (baud_en = '1') then
                               if (tx_bit_cnt = X"6") then tx_sm <= tx_stop;  end if;
                          end if;
         when tx_stop  => if (baud_en = '1') then        tx_sm <= tx_idle;    end if;
         when others => null;
       end case;
     end if;
   end process;
```

Fig. 5-15

1) The "baud" component, which we'll look at later. Note that we configure it with the BAUD_9600 constant from above;

2) the bit counter;

3) the state machine.

The rest of the code then implements the block functions shown above:

117

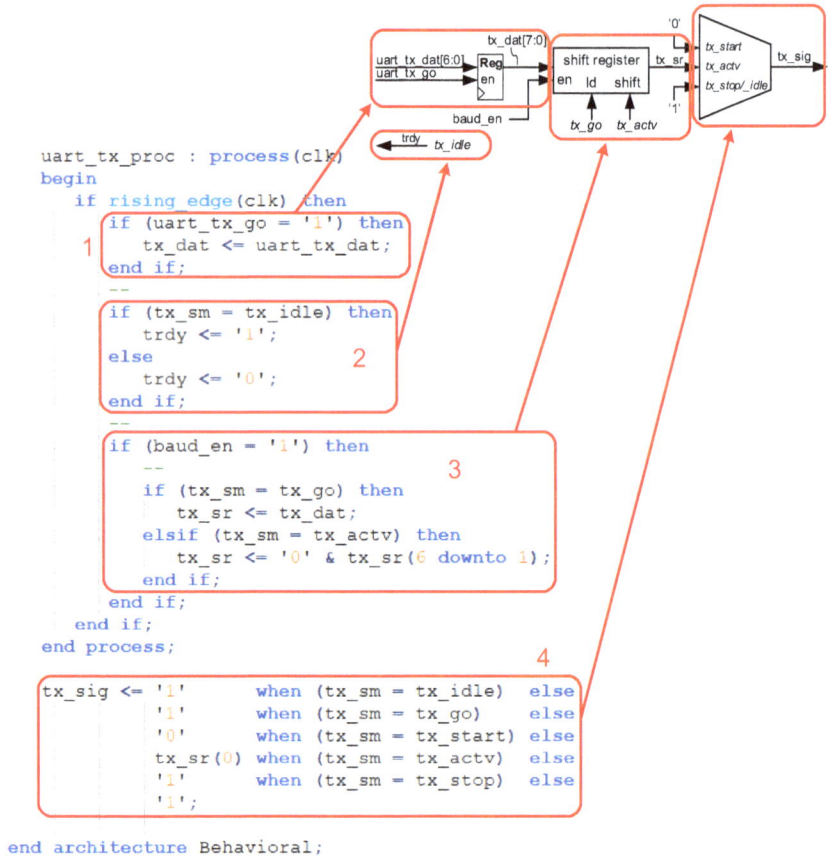

```
uart_tx_proc : process(clk)
begin
    if rising_edge(clk) then
        if (uart_tx_go = '1') then
            tx_dat <= uart_tx_dat;
        end if;
        --
        if (tx_sm = tx_idle) then
            trdy <= '1';
        else
            trdy <= '0';
        end if;
        --
        if (baud_en = '1') then
            --
            if (tx_sm = tx_go) then
                tx_sr <= tx_dat;
            elsif (tx_sm = tx_actv) then
                tx_sr <= '0' & tx_sr(6 downto 1);
            end if;
        end if;
    end if;
end process;

tx_sig <= '1'        when (tx_sm = tx_idle)  else
          '1'        when (tx_sm = tx_go)    else
          '0'        when (tx_sm = tx_start) else
          tx_sr(0)   when (tx_sm = tx_actv)  else
          '1'        when (tx_sm = tx_stop)  else
          '1';

end architecture Behavioral;
```

Fig. 5-16

1) The input data latch;

2) trdy—note that we register it instead of driving it directly from tx_idle state, since a direct combinatorial generation might create glitches on the trdy signal as the states change;

3) The loading and shifting of the shift register. Note how the shifting is done to the right (that line may require a bit of study—each clock, the bit 0 is replaced by the previous clock's bit 1), since a UART, unlike most other serial communication standards, sends the LS bit first. We again find the "&" concatenation symbol;

4) The combinatorial multiplexer. Most designs would include an output register, not so much to avoid state machine-changing glitches on the tx_sig output, but to avoid potential timing problems downstream. In general, it's good practice to register outputs.

This is a simulation of the UART Tx operation:

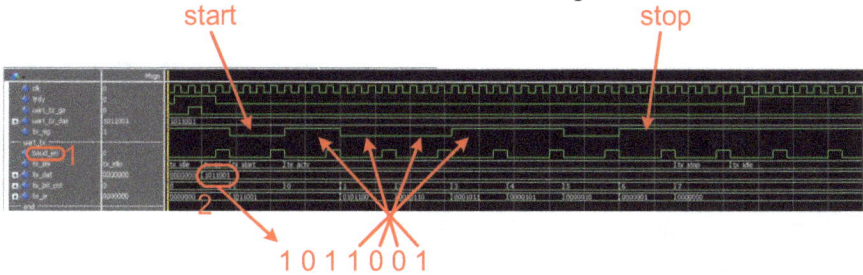

Fig. 5-17

1) Note that, for the sake of simulation, we've drastically shortened the BAUD period. Here, baud_en occurs every few 50 MHz clocks instead of every few thousand;
2) Note also that the sample binary value (0x59, or 1011001) is sent least-significant bit first.

UART, Rx

We next move on to the receiver half of the UART, which, as we'll see, comprises more than half the logic. We create the transmission bit by bit, but at the receiver side we must dissect the arriving signal to discern what the transmission side has wrought.

This is the basic concept for the UART receive operation:

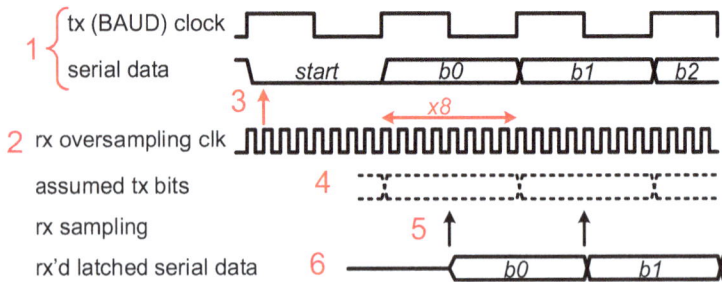

Fig. 5-18

1) The transmit side, as we've just covered;

2) On the receive side, we use a higher-speed clock, called an oversampling clock, to scan the received serial signal, looking for a start bit, i.e., simply looking for a low signal. Rx oversampling clocks are usually either eight or sixteen times the configured BAUD rate (and so, you can see that the oversampling clock would need to be configurable as well). For this example, we're using a x8 oversampling;

3) Once we find that the received signal has gone low, indicating the beginning of a start bit,

4) … we can establish where we think the original transmitted data occurs,

5) … and sample the received signal where we think the middle of each bit (the mid-bit) would be,

6) hopefully latching the same data that was transmitted.

We should note that this is the simplest approach possible. An actual implementation would likely include some amount of filtering for the start bit detection (so we don't get fooled by a line glitch), and re-calibration of the virtual received clock (and thus the mid-bit sampling point) with each subsequent data bit change (since an actual bit boundary is then manifested).

The oversampling clock is not actually a clock, at least not what we think of as a clock. In order to keep our design fully synchronous (always the goal), we use the original high-speed 50MHz clock whenever possible. Thus, the oversampling "clock" is actually an enable for the 50MHz clock that occurs at the oversampling rate.

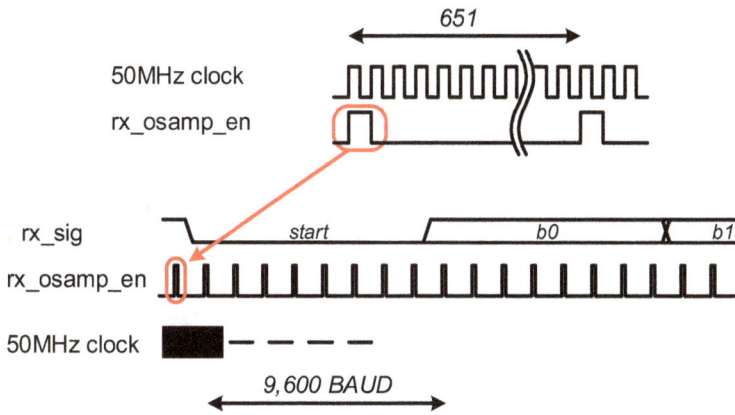

Fig. 5-19

At 50 MHz, the rx_osamp_en enable signal (Rx oversample enable) here repeats every 651 clocks, resulting in a repetition rate of $50 \times 10^6 / 651 = 76,805$ repetitions per second. Since there are eight enables per bit time, the final bit repetition is $76,805 / 8 = 9600.6$. Note that at the time scale of the bottom diagram, the 50 MHz clock is so dense that it appears as a solid bar.

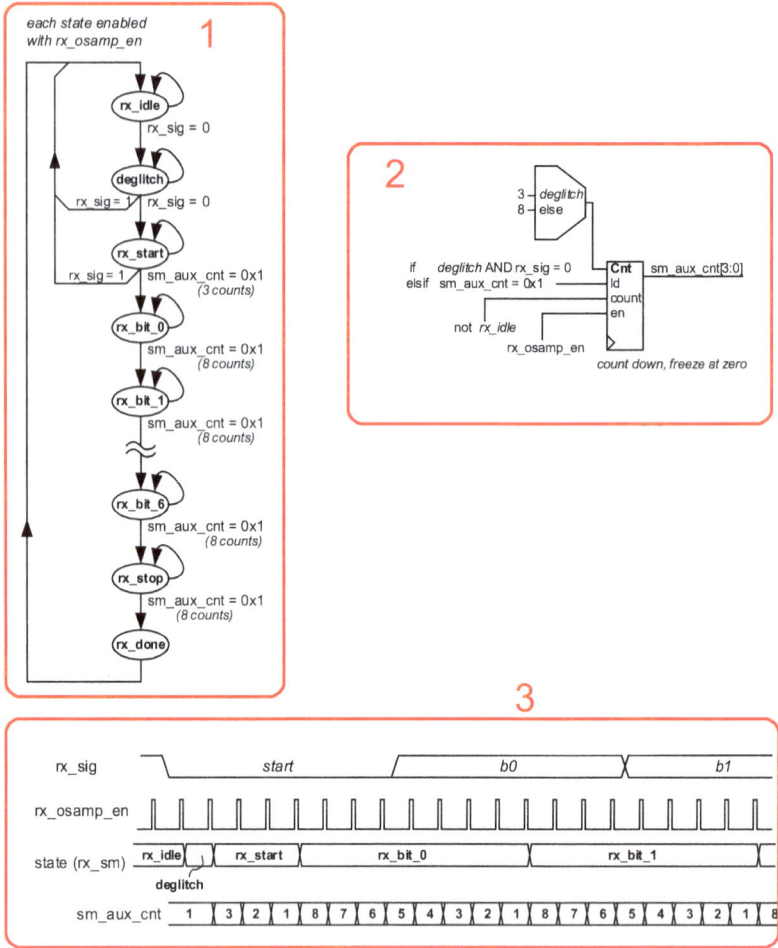

Fig. 5-20

Here's the state machine control:

1) The state machine is very linear. Once convinced that the start bit is indeed a bona fide start, the flow proceeds unconditionally through the transmission capture;

2) This is a counter that serves as an auxiliary resource for the state machine, as we'll soon see;

3) We'll refer to this timing diagram as we go along.

When in the "idle" state, any low occurrence of the received signal (at the instance of sampling signal rx_osamp_en) advances the state machine to the "deglitch" state.

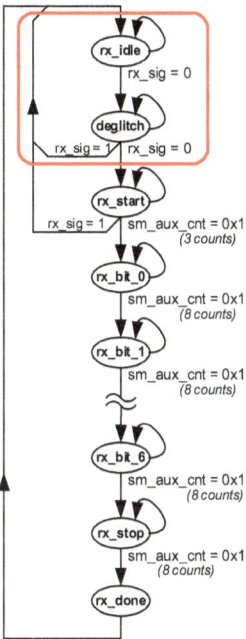

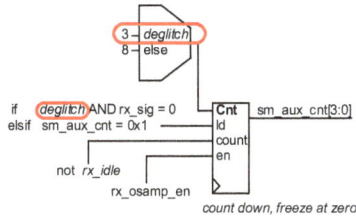

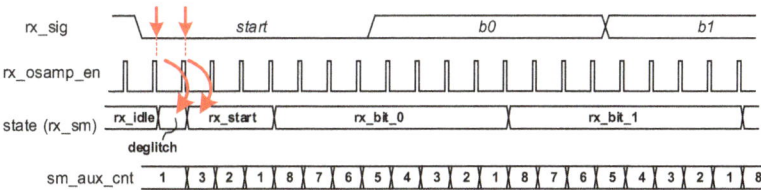

Fig. 5-21

The "deglitch" state makes sure that the observed low signal was not spurious noise. We also load our auxiliary counter with a value of three, the reason for which we'll see in a moment. If on the next rx_osamp_en the received signal is still low, we move on to the "rx_start" state.

The auxiliary counter allows us to program how long (how many rx_osamp_en sample times) the state machine waits in one state. Here we've loaded a value of three, so we wait for three rx_osamp_en occurrences before moving on. Instead of using an auxiliary counter, we could have let the state machine step through three individual states, but this expands the state machine with no functional advantage. Also, an auxiliary counter allows us to easily change the wait time if needed without re-coding the state machine.

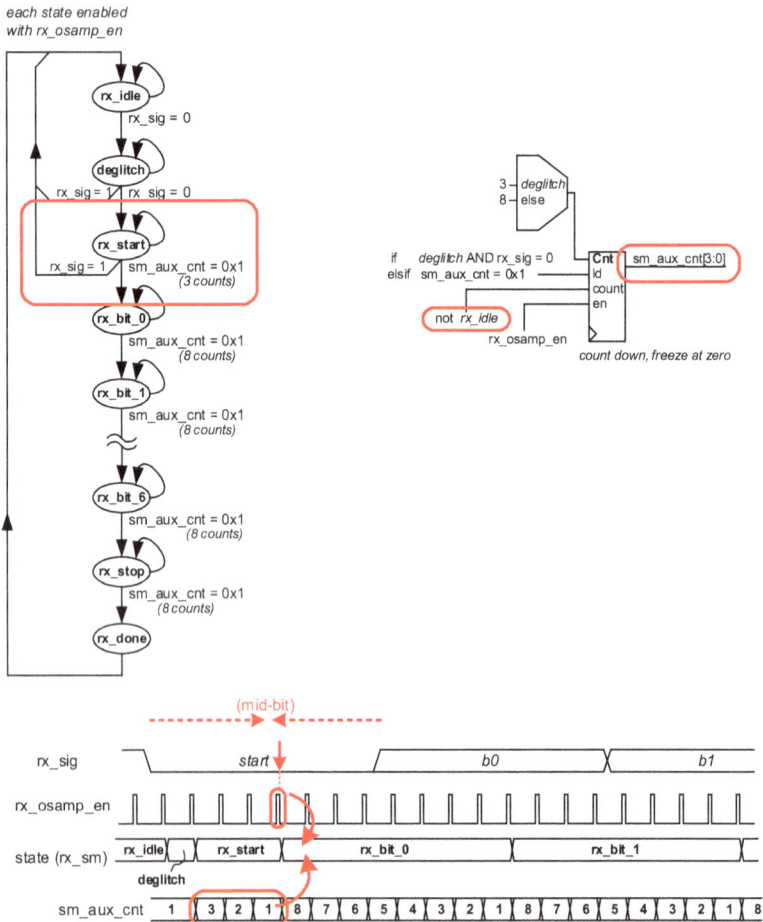

Fig. 5-22

Notice that by waiting three enable samples, the "rx_state" transitions to "rx_bit_0" halfway through the received signal start period. As we'll see, this establishes our mid-bit sampling.

Once the state machine leaves "rx_state", each received bit is associated with one state, and, since the auxiliary counter is now loaded each time with a value of eight, each subsequent state is also eight rx_osamp_en sample periods long.

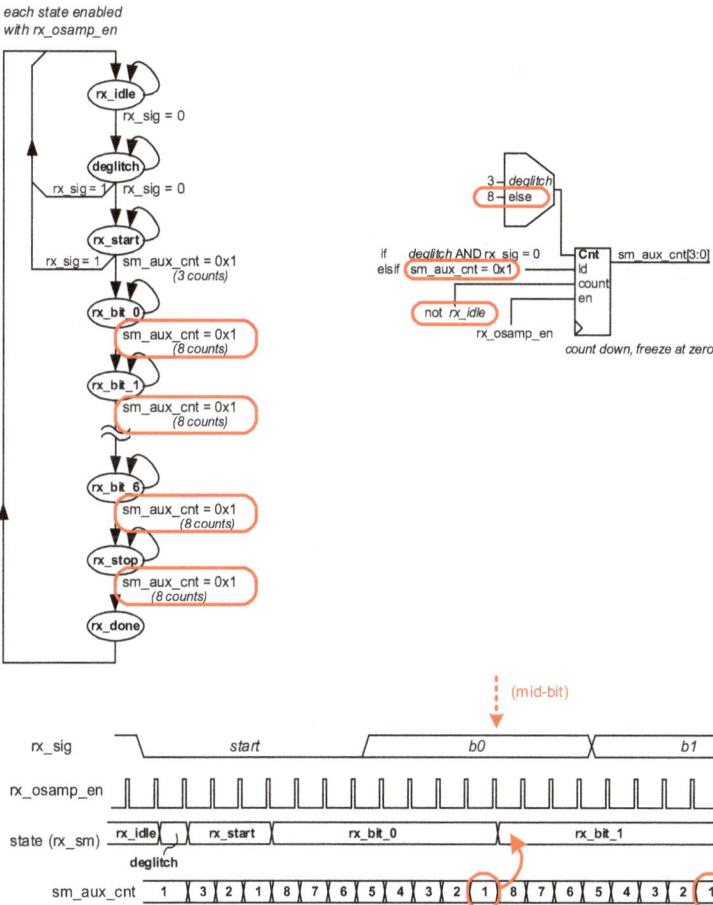

Fig. 5-23

The end of each bit state (e.g., "rx_bit_0") coincides with the mid-bit of its associated bit, and as we'll soon see, this is where we sample the received signal.

Now that we understand the basic control operation of the receive side of the UART, we next see how the bits are received and assembled into a seven-bit value and presented to, for example, software.

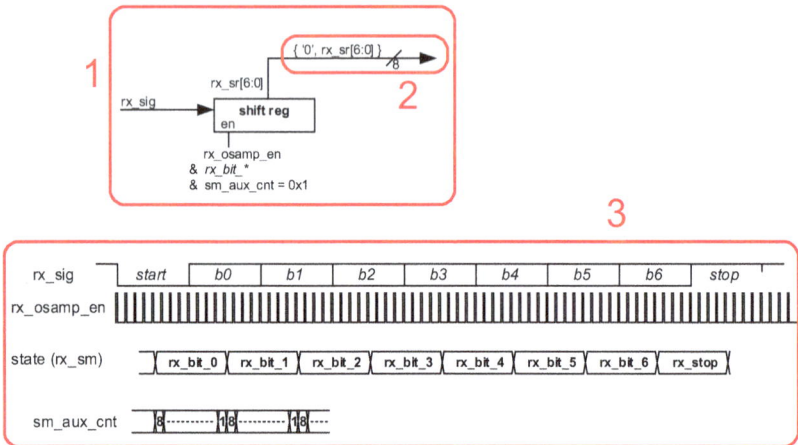

Fig. 5-24

1) Each sampled bit is placed into a shift register, which, after seven samples, results in a 7-bit parallel value, rx_sr[6:0],

2) which is then concatenated with an MSB (most-significant bit) zero to create an 8-bit byte (for easier handling later on);

3) This is the timing diagram we've already seen, except condensed to view the entire 7-bit assembly.

Next, we add the parallel output of the shift register:

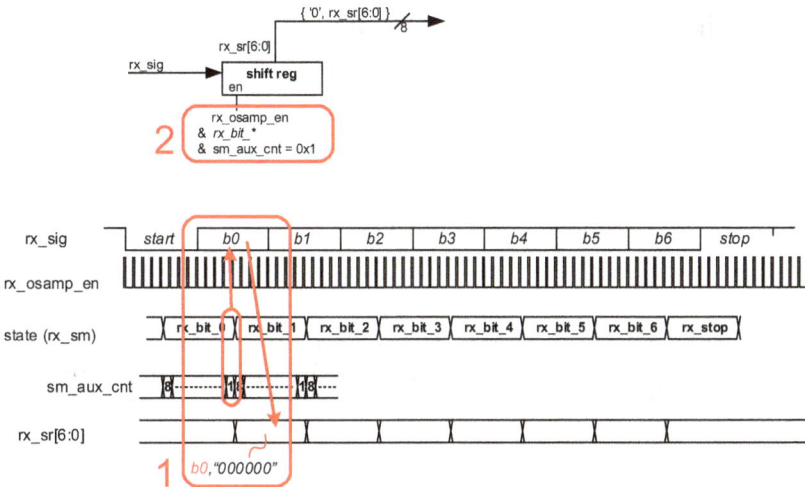

Fig. 5-25

1) Here, we see that when sm_aux_cnt reaches one, we enable the received rx_sig (bit 0 at this point in time) to be latched into the shift register. Note that bit 0 (b0) is located at the MSB of the shift register output, exactly opposite of what we want. As we'll soon see, this is just temporary;

2) The "*" asterisk is a wild-card in this block diagram, meaning that any of the (six) states beginning with "rx_bit_" can enable the shift register (noting that an asterisk is not used in the code, since it indicates multiplication in VHDL).

The next bit to be latched into the shift register is bit 1 (b1):

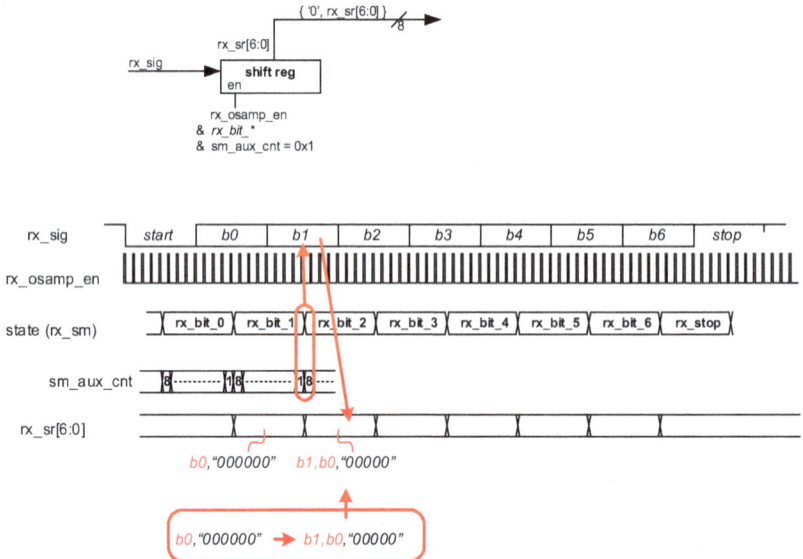

Fig. 5-26

Bit 1 replaces bit 0, which is now shifted to the right. Thus, this is a "shift-right" shift register.

The received bit shifting proceeds …

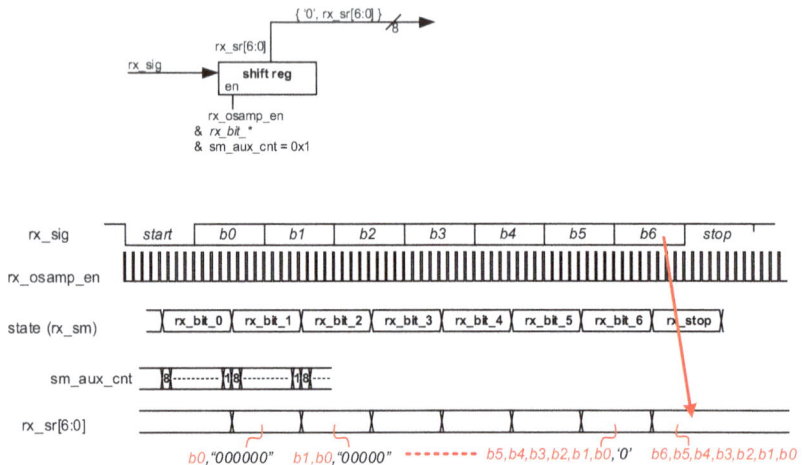

Fig. 5-27

... until the full seven-bit value is presented out of the shift register.

We have one more piece to add to the receive side of the UART. We assume that software is servicing the UART, meaning that the software is providing the UART Tx values on the transmit side, and accepting the UART Rx values on the receive side. And here's the rub: whereas on the transmit side, the software can choose to provide a Tx value at its discretion (based on the trdy status), it has no control (at least in our simple implementation) over when receive Rx values arrive. If the software is busy and fails to take a received value from the shift register before another comes along, the new one will overwrite the waiting one, and that one is then lost.

What we'd like is a mechanism to hold arriving values until the software is ready to take a look:

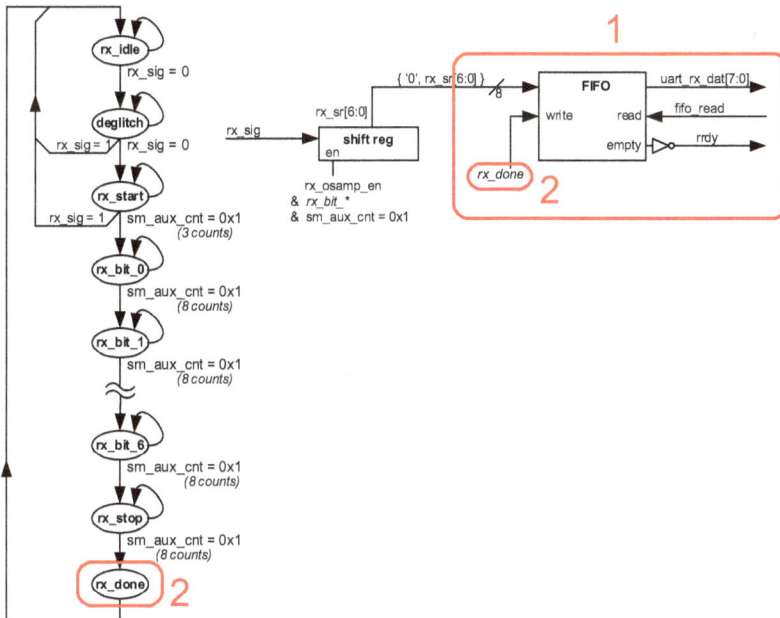

Fig. 5-28

1) The mechanism we'll use is a FIFO, which stands for "First In, First Out," and the operation is just as it sounds, much like people queuing at the theater or checkout counter, the customers are serviced in the same order that they arrive;

2) And now we understand the purpose of the one-clock "rx_done" state. Once an arriving receive value has been assembled in the shift register, we write it into the FIFO.

On the output side of the FIFO (where we're assuming the software operates):

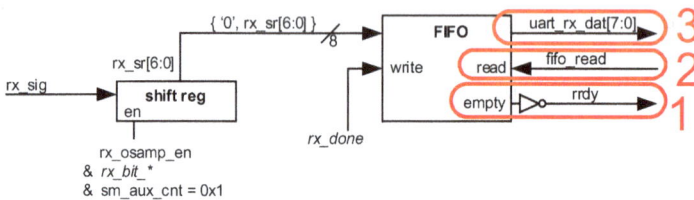

Fig. 5-29

1) The FIFO indicates when it's empty, and when it's not empty, i.e., it contains at least one value, the "rrdy" (Read Ready) flag goes active, letting the software know that received data is waiting;

2) The software pulses the "fifo_read" signal,

3) by which the FIFO presents the next data value (uart_rx_dat) in line to be presented. From here, the software would then check the "rrdy" line to see if any more values are available. Notice that the uart_rx_dat output data is eight bits wide. The user of the UART—the software—presumably understands that the LS seven bits comprise the received serial data. FIFOs are often constructed of multiples of byte widths (here just one), thus the concatenated eight bits of data.

UART, Rx, code

Next, we look at the VHDL code. The entity I/O matches that of the previous block diagram:

```vhdl
library IEEE;
use IEEE.STD_LOGIC_1164.all;
use IEEE.NUMERIC_STD.all;
use IEEE.STD_LOGIC_MISC.all;
use IEEE.STD_LOGIC_UNSIGNED.all;

entity uart_rx is
  port (
        clk            : in    std_logic;
        rx_sig         : in    std_logic;
        uart_rx_dat    : out   std_logic_vector(7 downto 0);
        fifo_read      : in    std_logic;
        rrdy           : out   std_logic
        );
end entity uart_rx;
```

Fig. 5-30

Next is the architecture declarations:

```
architecture Behavioral of uart_rx is

  -- baud constants assume a 50MHz clock
  constant BAUD_9600 : std_logic_vector(15 downto 0) := X"1458";

  component baud is
    port (
          clk           : in   std_logic;
          baud_config   : in   std_logic_vector(15 downto 0);
          baud_en       : out  std_logic
        );
  end component;                                                    1

  component fifo is
    port (
          clk           : in   std_logic;
          dat_in        : in   std_logic_vector(7 downto 0);
          f_write       : in   std_logic;
          dat_out       : out  std_logic_vector(7 downto 0);      2
          f_read        : in   std_logic;
          empty         : out  std_logic
        );
  end component;

  type rx_sm_type is ( rx_idle,
                       deglitch,
                       rx_start,
                       rx_bit_0,
                       rx_bit_1,
                       rx_bit_2,
                       rx_bit_3,          3
                       rx_bit_4,
                       rx_bit_5,
                       rx_bit_6,
                       rx_stop,
                       rx_done
                     );
  signal rx_sm : rx_sm_type;

  signal baud_div8       : std_logic_vector(15 downto 0);
  signal rx_osamp_en     : std_logic;
  signal sm_aux_cnt      : unsigned(3 downto 0) := X"0";
  signal bit_state       : std_logic;
  signal rx_sr           : std_logic_vector(6 downto 0) := "0000000";
  signal fifo_in         : std_logic_vector(7 downto 0) := X"00";
  signal fifo_write      : std_logic;
  signal fifo_empty      : std_logic;
```

Fig. 5-31

1) The Rx side uses the same BUAD-rate generator as the Tx, but, as we'll see, with an adaptation to create the times-eight rx_osamp_en oversampling signal. Notice, though, that the BAUD divisor, BAUD_9600, is the same as was used in the Tx half;

2) The FIFO as shown in the earlier diagram. We can't name the write and read signals as simply "write" and "read", since these are reserved VHDL key words;

3) The state machine labels as shown in the earlier diagram.

Moving on to the architecture body, we begin with the BAUD generator:

```
begin

  --shift right x3 (by 8 divide)
  baud_div8 <= "000" & BAUD_9600(15 downto 3);

  baud_inst : baud
    port map
      (
        clk            => clk,        --in  std_logic;
        baud_config    => baud_div8,  --in  std_logic_vector(15 downto 0);
        baud_en        => rx_osamp_en --out std_logic
      );
```

Fig. 5-32

Here, we see that the BAUD divisor, BAUD_9600, is divided by eight. By discarding the LS three bits, we can see that the result is 1/8 of the original. Further, by decreasing the BAUD divisor, we increase the frequency—eight times more baud_en occurrences per unit time.

Next is the state machine, along with its attending auxiliary counter:

```
uart_rx_sm_aux_cnt : process(clk)
begin
  if rising_edge(clk) then                                  4
    if (rx_osamp_en = '1') then
      if (    rx_sm = deglitch                              3
          AND rx_sig = '0'
         ) then                                             5
        sm_aux_cnt <= X"3";
      elsif (sm_aux_cnt = X"1") then                        6
        sm_aux_cnt <= X"8";
      elsif (rx_sm /= rx_idle) then                         7
        sm_aux_cnt <= sm_aux_cnt - 1;
      end if;
    end if;
  end if;
end process;                                                                  1

uart_rx_sm : process(clk)
begin
  if rising_edge(clk) then                                  2
    if (rx_osamp_en = '1') then
      case (rx_sm) is
        when rx_idle   => if (rx_sig = '0') then                    rx_sm <= deglitch;   end if;
        when deglitch  => if (rx_sig = '1') then                    rx_sm <= rx_idle;
                          else                                      rx_sm <= rx_start;   end if;
        when rx_start  => if (rx_sig = '1') then                    rx_sm <= rx_idle;
                          elsif (sm_aux_cnt = X"1") then            rx_sm <= rx_bit_0;   end if;
        when rx_bit_0  => if (sm_aux_cnt = X"1") then               rx_sm <= rx_bit_1;   end if;
        when rx_bit_1  => if (sm_aux_cnt = X"1") then               rx_sm <= rx_bit_2;   end if;
        when rx_bit_2  => if (sm_aux_cnt = X"1") then               rx_sm <= rx_bit_3;   end if;
        when rx_bit_3  => if (sm_aux_cnt = X"1") then               rx_sm <= rx_bit_4;   end if;
        when rx_bit_4  => if (sm_aux_cnt = X"1") then               rx_sm <= rx_bit_5;   end if;
        when rx_bit_5  => if (sm_aux_cnt = X"1") then               rx_sm <= rx_bit_6;   end if;
        when rx_bit_6  => if (sm_aux_cnt = X"1") then               rx_sm <= rx_stop;    end if;
        when rx_stop   => if (sm_aux_cnt = X"1") then               rx_sm <= rx_done;    end if;
        when rx_done   =>                                           rx_sm <= rx_idle;
        when others => null;
      end case;
    end if;
  end if;
end process;
```

Fig. 5-33

1) The Rx state machine,

2) enabled by the oversampling signal;

3) The auxiliary counter,

4) also enabled by the oversampling signal;

5) From the deglitch state, the counter is loaded with a value of three so that it steps out three counts (during the rx_start state), locating the sequence halfway through the first data bit,

6) and from there, it's loaded with a value of eight, counting eight steps to move ahead to the middle of the next data bit;

7) Since the auxiliary counter is continually in operation, it counts down from whatever has been loaded (except when idle between UART transmissions).

Next, we look at the Rx shift register:

```
bit_state <= '1' when (   rx_sm = rx_bit_0
                       OR rx_sm = rx_bit_1
                       OR rx_sm = rx_bit_2
                       OR rx_sm = rx_bit_3
                       OR rx_sm = rx_bit_4          1
                       OR rx_sm = rx_bit_5
                       OR rx_sm = rx_bit_6
                     )
                 else
             '0';

shift_register : process(clk)
begin
   if rising_edge(clk) then                3
      if (rx_osamp_en = '1') then
         if (    (bit_state = '1')
             AND (sm_aux_cnt = X"1")     4
             ) then                              2
            rx_sr <= rx_sig & rx_sr(6 downto 1);
         end if;
      end if;
   end if;
end process;
```

Fig. 5-34

1) The shift register advances with each of the seven arriving bits of the UART data value, associated with seven "bit" states. Here, we create an intermediate signal, big_state, simply to keep the subsequent shift register operation clearer;

2) And we shift the bit,

3) as gated by our oversample enable,

4) whenever we're sitting in one of our associated "bit" states, and the auxiliary counter reaches its termination count (one).

Finally, we have the FIFO:

```
fifo_in    <= ('0' & rx_sr);                                    1
fifo_write <= '1' when (rx_osamp_en = '1' AND rx_sm = rx_done)  2
                    else
              '0';

rx_fifo : fifo                                          3
   port map
   (
      clk     => clk,       --in    std_logic;
      dat_in  => fifo_in,   --in    std_logic_vector(7 downto 0);
      f_write => fifo_write, --in   std_logic;
      dat_out => uart_rx_dat, --out std_logic_vector(7 downto 0);
      f_read  => fifo_read,  --in   std_logic;
      empty   => fifo_empty  --out  std_logic
   );

   rrdy <= NOT fifo_empty;  4

end architecture Behavioral;
```

Fig. 5-35

1) We expand the FIFO input to eight bits by concatenating an MS zero,

2) and create a FIFO write signal when the state machine reaches the end—VHDL is not very helpful about including this sort of logic directly in the port map assignments (unlike Verilog);

3) The FIFO instantiation, per the diagram,

4) and inverting the FIFO empty output in order to create a true-active read-ready flag.

FIFOs are generally provided by the FPGA vendors as IP (Intellectual Property) modules, and accessed and used within the vendor's compiler software. Vendor-supplied functional IP modules have the great advantage that you don't have to design them yourself, and they typically include more functionality than your patience (or employer's money) would allow. Also, they come fully tested and documented. The minor downside is that they are not transportable, meaning that if you move your design to a different vendor, you'll have to re-generate the module.

Here's a very simple FIFO suitable for simulating the UART design:

```
architecture Behavioral of fifo is                              1

    type memory_type is array(0 to 7) of std_logic_vector(7 downto 0);
    signal memory : memory_type;

    signal write_adr          : unsigned(2 downto 0) := "000"; 3
    signal read_adr           : unsigned(2 downto 0) := "000";
    signal write_adr_fifo     : integer;
    signal read_adr_fifo      : integer;

begin

    fifo_addresses : process(clk)              2
    begin
       if rising_edge(clk) then
          if (f_write = '1') then
             write_adr <= write_adr + 1;
          end if;
          --
          if (f_read = '1') then
             read_adr <= read_adr + 1;
          end if;
       end if;
    end process;

    write_adr_fifo <= to_integer(write_adr + 1);  3
    read_adr_fifo  <= to_integer(read_adr);

    memory_process : process(clk)
    begin
       if rising_edge(clk) then
          if (f_write = '1') then
             memory(write_adr_fifo) <= dat_in;    4
          end if;
          --
          dat_out <= memory(read_adr_fifo);
       end if;
    end process;                               5

    empty <= '1' when (write_adr = read_adr) else '0';

end architecture Behavioral;
```

Fig. 5-36

1) A FIFO is, at core, a dual-port memory, with a write capability on one side (the input), and a read capability on the other (the output). In VHDL, a memory is usually coded using a one-dimensional array, such as you would encounter in any software programming language. A VHDL array is declared using the keyword "array", followed by the depth, here a depth of eight (0 to 7). Our memory array consists of std_logic_vector types, although other types are allowed;

2) Each write flag occurrence increments the input address, and each read flag increments the output address;

3) Although the array contents can be a variety of types (besides our std_logic_vector), indexing the array—pointing to a content location—must be done with integers. Since our addresses are the array indexes, we must convert the unsigned arithmetic-capable versions to integers, and this is done with the "to_integer" keyword. Note that, in the process, we increment the write address—this is so that our FIFO does not operate as a "fall-through" type, meaning that our FIFO output value is accessed via the read command (f_read). If the FIFO were a fall-through type, the written value would appear immediately at the output (assuming the FIFO were empty at the time). Fall-through FIFOs have their uses, but require special care to keep the operation synchronized on the input and output;

4) This process implements the dual-port memory. Since ours is not read/write on both sides, the write and read sides are essentially one line each. Note that, whereas the input data is written using the f_write command signal (which also increments the write address in preparation for a subsequent write), the contents of the read side are continually present—the read command, f_read, simply advances the read address;

5) The FIFO is considered empty whenever the write and read addresses are the same (as at the beginning, before we write the first UART value), since whenever there are unread values, the write address will be at least one more than the read.

As indicated above, this is the simplest FIFO possible. We've made it small, just 8 words deep. When more than eight words are entered, the write address counter rolls over—meaning that the counter returns to zero and counts up again. This is fine, since the read address following behind does the same. Although we've included an empty flag indication, most FIFOs also provide a full indication, or even a configurable partial-full flag. Our empty flag is easily implemented—just detecting that the write and read counter addresses are the same. You can see that a full or partial-full flag requires a good deal more arithmetic and logic, since, once the write address rolls over, the read address is greater than the write address, at least until the read address also rolls over. And without a full or partial-full flag, the write side might exceed the FIFO depth, resulting in all manner of chaos.

Here's a simulation of three transmitted UART words, where the output of the Tx UART is connected to the input of the Rx, and, like the Tx UART simulation above, we've drastically shortened the timescale (e.g., rx_osamp_en occurs every other clock, instead of every 651 for 9600 BAUD):

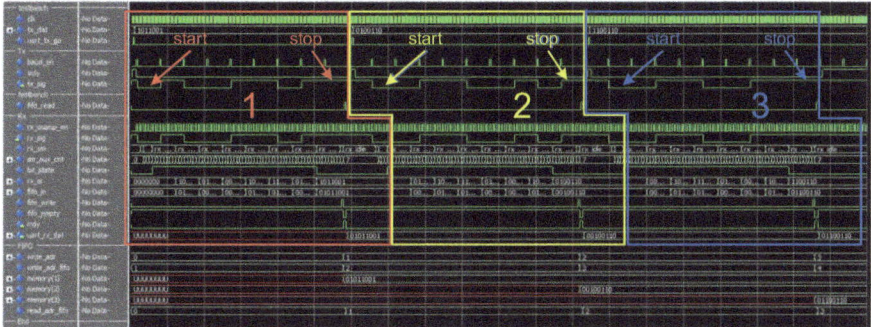

Fig. 5-37

1) The first transmitted and received UART word;
2) The second transmitted and received UART word;
3) The third transmitted and received UART word;

Note that in the simulation, the receiver of the Rx side (e.g., software) accepts each received word (reads the FIFO) immediately.

We'll Zoom in on the second word:

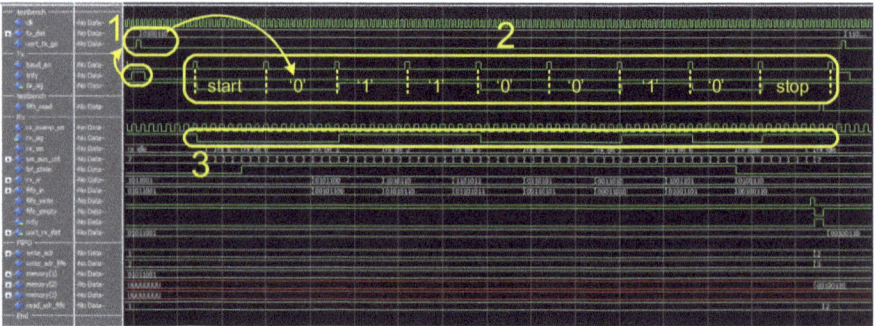

Fig. 5-38

1) The Tx UART indicates that it's ready to transmit a word, and the testbench responds by presenting a seven-bit "0100110" value and activating the uart_tx_go signal command,

2) whereby the Tx UART serializes the data and sends it out on the serial output,

3) and, since we've tied the Tx UART output to the Rx input, that is the same.

The Rx UART assembles (de-serializes) the arriving signal, recovers the original "0100110" value,

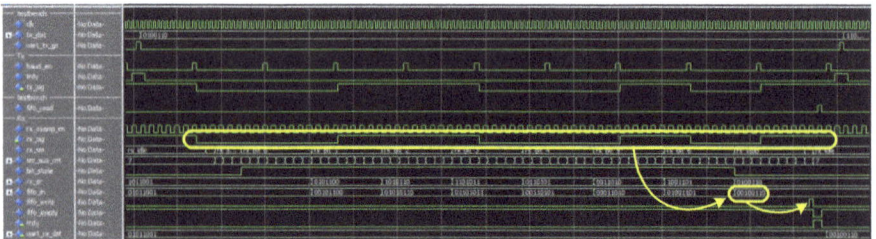

Fig. 5-39

… and writes this into the FIFO (via fifo_write).

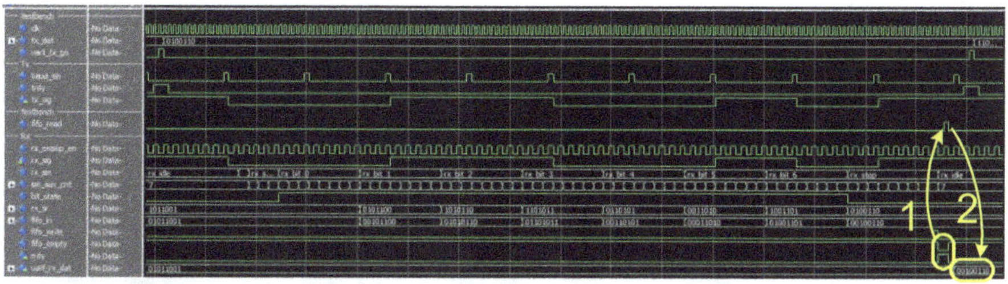

Fig. 5-40

1) The Rx UART translates the resulting FIFO empty flag into rrdy,

2) and the testbench, seeing this, then reads the FIFO (via fifo_read) to obtain the recovered data.

Rather than software immediately taking a received UART word (by reading the FIFO), a more likely scenario would be that software would be busy on other tasks, and only occasionally turn to take the received words. The following simulation shows three received words queued in the FIFO before the testbench (emulating the software) reads them.

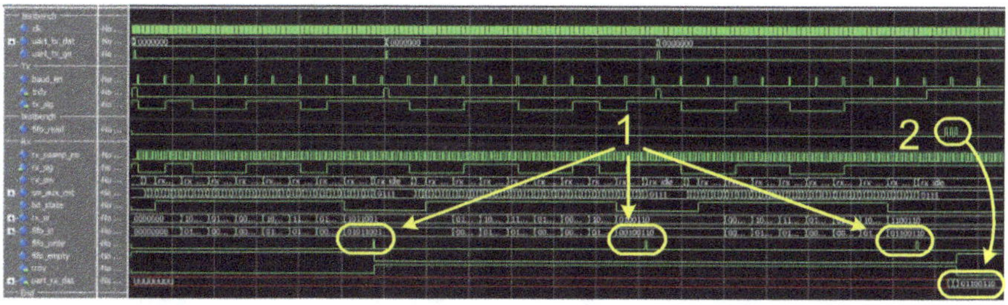

Fig. 5-41

1) We see the same writes into the FIFO as before,
2) but the testbench waits to read them.

Here we zoom in on the testbench reads:

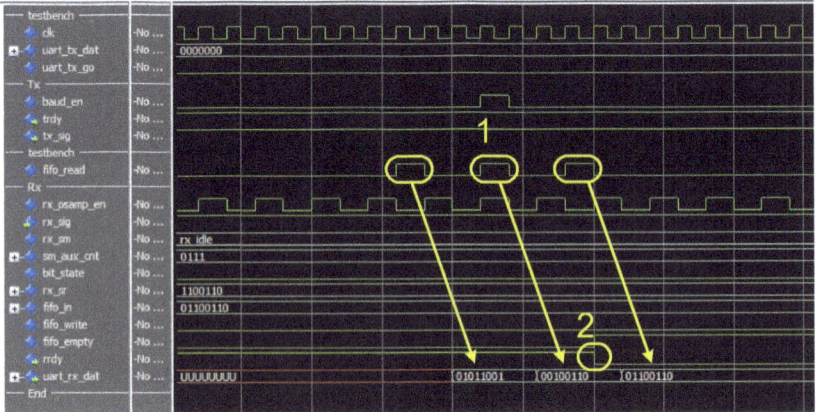

Fig. 5-42

1) The testbench performs three consecutive FIFO reads;
2) and then stops as the "ready" flag goes inactive.

Chapter 6

Memory-mapped Buses

We've talked about software "servicing" the user side of the Tx and Rx UARTs, and we'll now look at a common means to do so, the memory-mapped bus. These buses are termed "memory-mapped" because from the software's perspective, everything it writes to or reads from on the bus appears to be part of one big memory, where each register is accessed via a unique address on the bus. Of course, a sub-set of the bus addresses could write/read an actual memory block.

Here's a very simple example:

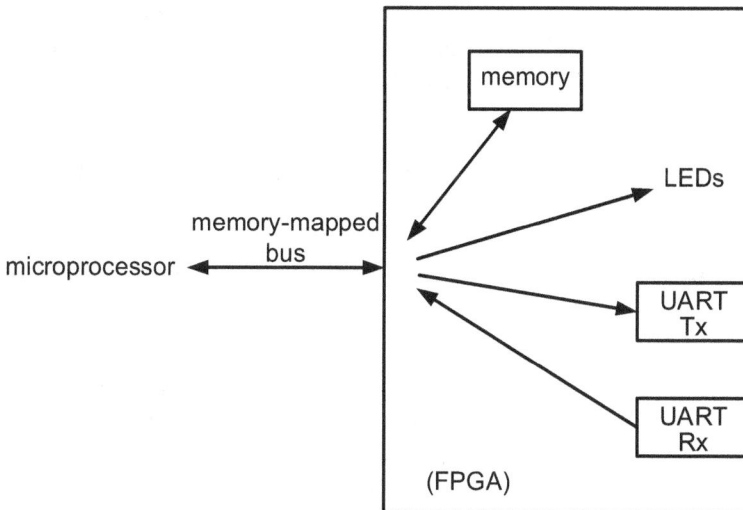

Fig. 6-1

In addition to our Tx and Rx UARTS, this memory-mapped bus writes to some LEDs, and accesses a memory block (usually provided by the FPGA vendor as configured IP).

This is a break-out of the bus:

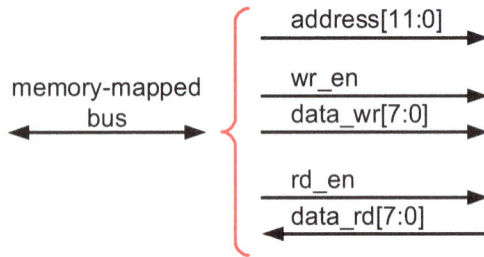

Fig. 6-2

The "address" selects among the registers (e.g., LEDs, Tx UART data, etc.) as well as the individual locations of the memory block.

The "wr_en" is the write enable, and the value present on the "data_wr" signal is latched (or written, in the case of the memory) to the addressed location when "wr_en" is active.

The "rd_en" instructs the addressed register (or memory location) to place its value on the "data_rd" line when "rd_en" is active.

Note that the write and read data signals are eight bits (a byte). This is all that's needed for the LEDs and our UARTs, but often the FPGA operation includes much more complex operations, and uses data buses that are 32 or even 128 or more bits wide.

Also, with an address signal of twelve bits, we can access 2^{12} locations, or 4K (4096). Most address signals in an actual application would be far wider than this. After all, gigabit memories are common now, and one gigabit requires at least 31 bits.

This is the address allocations for our very simple example:

address	target
0x000	Tx UART FIFO ready status
0x001	Tx UART "go" command
0x002	Tx UART data
0x003	Rx UART FIFO not-empty status
0x004	Rx UART "read" command
0x005	Rx UART data
0x006	LEDs
0x007 - 0x7FF	(not used)
0x800 - 0xFFF	memory (2K)

Fig. 6-3

Note that we're using half the address space for our 2K (2048) memory.

Memory-mapped Bus Interfaces

For writes, when the microprocessor activates the write enable signal (wr_en), we must pass that along, with the data to be written (data_wr[7:0]), to the proper addressed target.

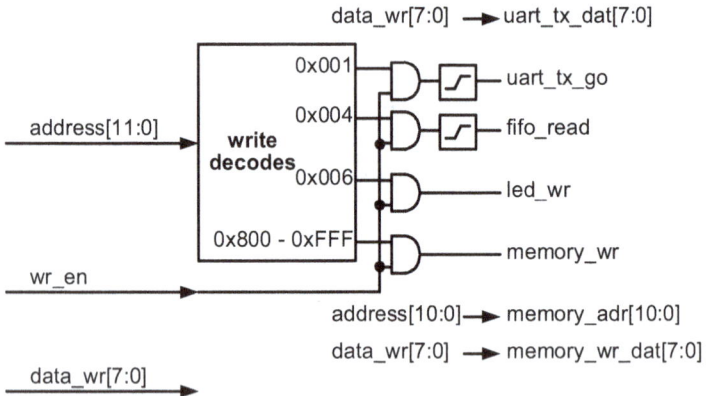

Fig. 6-4

Here's the breakdown:

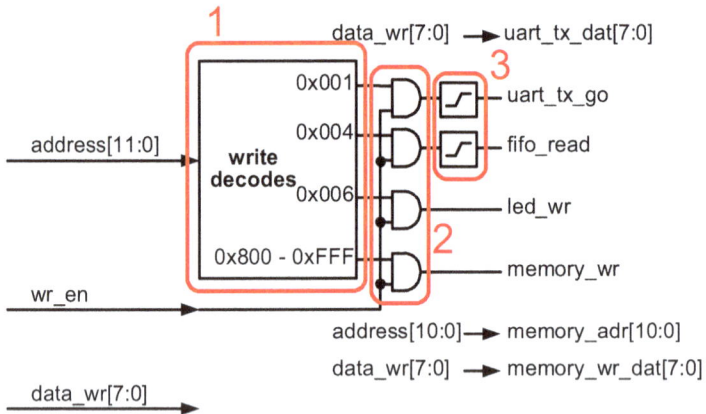

Fig. 6-5

1) The outputs of the decoder go high when their specific address (or addresses, in the case of the memory) are presented;

2) These then gate the write enable signal for each target;

3) The UARTs expect control signals to be one clock wide, so we must limit their gated write enables to just that, via the rising-edge detectors we saw earlier.

Read requests select one of the read sources to be placed on the processor's memory-mapped read signal.

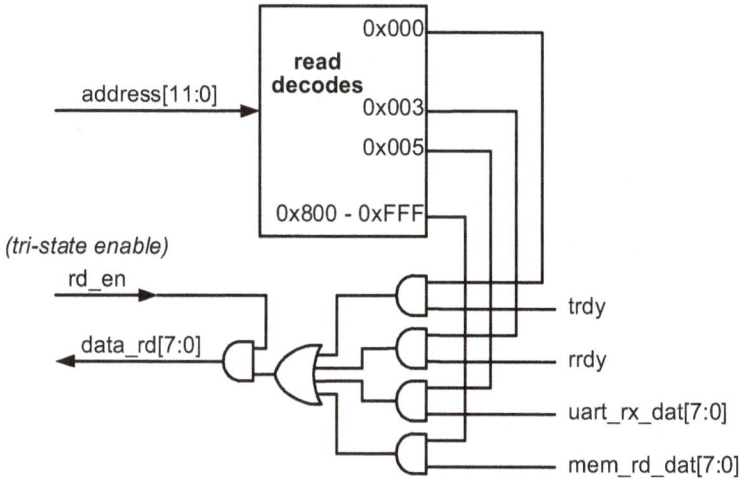

Fig. 6-6

And, the operation is as follows:

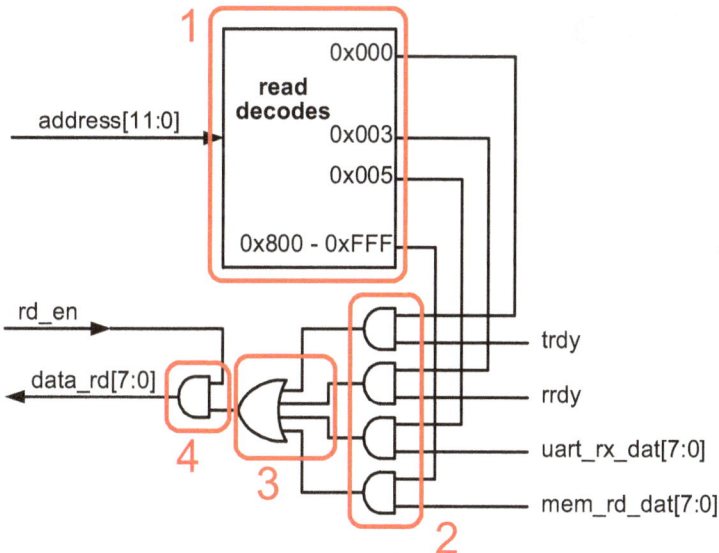

Fig. 6-7

1) The read decodes work exactly the same as those of the writes;

2) Each decode output enables the associated read source to pass through, otherwise that source is forced to zero;

3) These are then OR'd, noting that, since all but one is zero, the OR gate output is the same as the selected read source;

4) Finally, the output to the processor on the memory-mapped bus is gated with the read enable signal. In a very simple example, where this is the only read source, this enable is not necessary. If other additional sources come together to share the same memory-mapped read signal, however, then this AND gate serves the same purpose as those of 2) above.

We use logic gating here to multiplex the various read sources together, and we should note that another method (e.g., external to the FPGA) could use tri-state drivers that serve the same purpose (but instead of non-selected sources being forced to zero, there they would be held in an undriven tri-state mode so that the selected source could drive the line).

Example Design With a Shell

We'll now collect the various functional blocks (UARTs, LED, memory) together, along with the processor memory-mapped bus interface, in a high-level file that we call a shell, since its main purpose is to simply house and interconnect these various blocks.

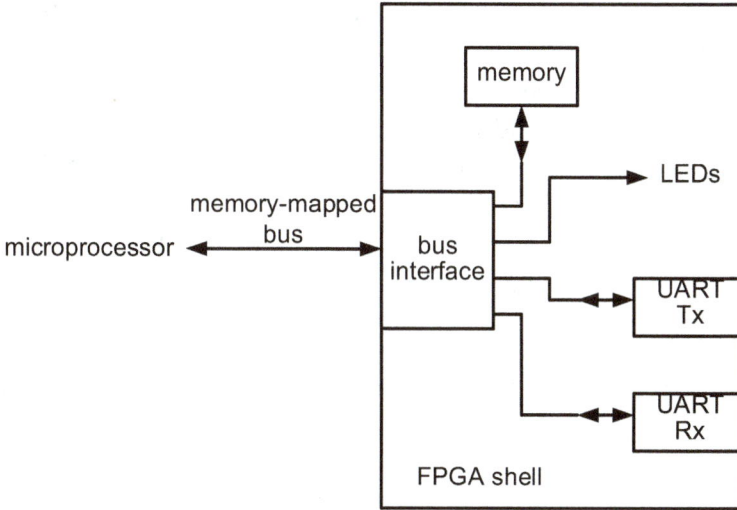

Fig. 6-8

This is the VHDL entity declaration for this shell file:

```
--  -------------------------------------------------------
--   (Useful information.)
--  -------------------------------------------------------

library IEEE;
use IEEE.STD_LOGIC_1164.all;
use IEEE.NUMERIC_STD.all;
use IEEE.STD_LOGIC_MISC.all;
use IEEE.STD_LOGIC_UNSIGNED.all;

entity shell is
                                    1
  generic (BAUD_9600 : std_logic_vector(15 downto 0));
  port (
        clk          : in    std_logic;
        address      : in    std_logic_vector(11 downto 0);
   2    wr_en        : in    std_logic;
        data_wr      : in    std_logic_vector(7 downto 0);
        rd_en        : in    std_logic;
        data_rd      : out   std_logic_vector(7 downto 0);
        --
        uart_tx_sig  : out   std_logic;
   3    uart_rx_sig  : in    std_logic;
        leds         : out   std_logic_vector(3 downto 0)
        );
end entity shell;
```

Fig. 6-9

1) We're using a generic to provide the overall UART 9600 BAUD rate. This implies that there is at least one VHDL module above this one in the hierarchy—in this case, a testbench (as we'll soon see), allowing a modified BAUD rate for simulation. In an actual application, this generic would probably provide a general BAUD value (versus specific for 9600 operation);
2) The memory-mapped bus;
3) External IO.

Here's the module declaration of the new Bus Interface module:

```
architecture Behavioral of shell is

  component bus_intfc is
    port (
          clk           : in    std_logic;
          address       : in    std_logic_vector(11 downto 0);
          wr_en         : in    std_logic;
          data_wr       : in    std_logic_vector(7 downto 0);
          rd_en         : in    std_logic;
          data_rd       : out   std_logic_vector(7 downto 0);
          --
          trdy          : in    std_logic;
          uart_tx_go    : out   std_logic;
          uart_tx_dat   : out   std_logic_vector(7 downto 0);
          --
          rrdy          : in    std_logic;
          fifo_read     : out   std_logic;
          uart_rx_dat   : in    std_logic_vector(7 downto 0);
          --
          leds          : out   std_logic_vector(3 downto 0);
          --
          memory_adr    : out   std_logic_vector(10 downto 0);
          memory_wr     : out   std_logic;
          memory_wr_dat : out   std_logic_vector(7 downto 0);
          memory_rd_dat : in    std_logic_vector(7 downto 0)
        );
  end component;
```

Fig. 6-10

1) The memory-mapped bus (the "user" side of the Bus Interface);
2) Tx UART;
3) Rx UART;
4) The internal memory interface.

We'll cover the design of this module soon.

The rest of the module declarations, and the signal declarations:

```
component uart_tx is
  generic (BAUD_9600 : std_logic_vector(15 downto 0));
  port (
       clk           : in    std_logic;
       uart_tx_dat   : in    std_logic_vector(6 downto 0);
       uart_tx_go    : in    std_logic;
       tx_sig        : out   std_logic;
       trdy          : out   std_logic := '0'
       );
end component;

component uart_rx is
  generic (BAUD_9600 : std_logic_vector(15 downto 0));
  port (
       clk           : in    std_logic;
       rx_sig        : in    std_logic;
       uart_rx_dat   : out   std_logic_vector(7 downto 0);
       fifo_read     : in    std_logic;
       rrdy          : out   std_logic
       );
end component;

component memory is
  port (
       clk           : in    std_logic;
       memory_adr    : in    std_logic_vector(10 downto 0);
       memory_wr     : in    std_logic;
       memory_wr_dat : in    std_logic_vector(7 downto 0);
       memory_rd_dat : out   std_logic_vector(7 downto 0)
       );
end component;

signal uart_tx_go    : std_logic;
signal uart_tx_dat_8 : std_logic_vector(7 downto 0);
signal uart_rx_dat   : std_logic_vector(7 downto 0);
signal trdy          : std_logic := '0';
signal rrdy          : std_logic := '0';
signal fifo_read     : std_logic;
signal memory_adr    : std_logic_vector(10 downto 0);
signal memory_wr     : std_logic;
signal memory_wr_dat : std_logic_vector(7 downto 0);
signal memory_rd_dat : std_logic_vector(7 downto 0);

begin
```

Fig. 6-11

Note that these signals simply inter-connect the instantiated modules.

And, finally, the module instantiations:

```vhdl
begin

   bus_intfc_inst : bus_intfc
      port map
      (
         clk            => clk,           --in   std_logic;
         address        => address,       --in   std_logic_vector(11 downto 0);
         wr_en          => wr_en,         --in   std_logic;
         data_wr        => data_wr,       --in   std_logic_vector(7 downto 0);
         rd_en          => rd_en,         --in   std_logic;
         data_rd        => data_rd,       --out  std_logic_vector(7 downto 0);

         trdy           => trdy,          --in   std_logic;
         uart_tx_go     => uart_tx_go,    --out  std_logic;
         uart_tx_dat    => uart_tx_dat_8, --out  std_logic_vector(7 downto 0);
         --
         rrdy           => rrdy,          --in   std_logic;
         fifo_read      => fifo_read,     --out  std_logic;
         uart_rx_dat    => uart_rx_dat,   --in   std_logic_vector(7 downto 0);
         --
         leds           => leds,          --out  std_logic_vector(3 downto 0);
         --
         memory_adr     => memory_adr,    --out  std_logic_vector(10 downto 0);
         memory_wr      => memory_wr,     --out  std_logic;
         memory_wr_dat  => memory_wr_dat, --out  std_logic_vector(7 downto 0);
         memory_rd_dat  => memory_rd_dat  --in   std_logic_vector(7 downto 0)
      );

   uart_tx_inst : uart_tx
      generic map (BAUD_9600 => BAUD_9600)
      port map
      (
         clk            => clk,                        --in   std_logic;
         uart_tx_dat    => uart_tx_dat_8(6 downto 0),  --in   std_logic_vector(6 downto 0);
         uart_tx_go     => uart_tx_go,                 --in   std_logic;
         tx_sig         => uart_tx_sig,                --out  std_logic;
         trdy           => trdy                        --out  std_logic := '0'
      );

   uart_rx_inst : uart_rx
      generic map (BAUD_9600 => BAUD_9600)
      port map
      (
         clk            => clk,           --in   std_logic;
         rx_sig         => uart_rx_sig,   --in   std_logic;
         uart_rx_dat    => uart_rx_dat,   --out  std_logic_vector(7 downto 0);
         fifo_read      => fifo_read,     --in   std_logic;
         rrdy           => rrdy           --out  std_logic
      );

   memory_inst : memory
      port map
      (
         clk            => clk,           --in   std_logic;
         memory_adr     => memory_adr,    --in   std_logic_vector(10 downto 0);
         memory_wr      => memory_wr,     --in   std_logic;
         memory_wr_dat  => memory_wr_dat, --in   std_logic_vector(7 downto 0);
         memory_rd_dat  => memory_rd_dat  --out  std_logic_vector(7 downto 0)
      );

end architecture Behavioral;
```

Fig. 6-12

1) External signals (in/out of the shell module);
2) A few interconnecting signals.

As you can see, this shell file simply instantiates and interconnects the various modules of the design. We previously covered the UART (Tx and Rx) and memory modules, and next, we'll look at the design of the Bus Interface module.

The Bus Interface module entity declaration, with the I/O signals we've just seen:

```vhdl
-- ------------------------------------------------------
--   (Useful information.)
-- ------------------------------------------------------

library IEEE;
use IEEE.STD_LOGIC_1164.all;
use IEEE.NUMERIC_STD.all;
use IEEE.STD_LOGIC_MISC.all;
use IEEE.STD_LOGIC_UNSIGNED.all;

entity bus_intfc is
  port (
          clk           : in    std_logic;
          address       : in    std_logic_vector(11 downto 0);
          wr_en         : in    std_logic;
          data_wr       : in    std_logic_vector(7 downto 0);
          rd_en         : in    std_logic;
          data_rd       : out   std_logic_vector(7 downto 0);
          --
          trdy          : in    std_logic;
          uart_tx_go    : out   std_logic;
          uart_tx_dat   : out   std_logic_vector(7 downto 0);
          --
          rrdy          : in    std_logic;
          fifo_read     : out   std_logic;
          uart_rx_dat   : in    std_logic_vector(7 downto 0);
          --
          leds          : out   std_logic_vector(3 downto 0);
          --
          memory_adr    : out   std_logic_vector(10 downto 0);
          memory_wr     : out   std_logic;
          memory_wr_dat : out   std_logic_vector(7 downto 0);
          memory_rd_dat : in    std_logic_vector(7 downto 0)
      );
end entity bus_intfc;

architecture Behavioral of bus_intfc is
```

Fig. 6-13

... and this is the signal instantiations, and the Bus Interface "writes" operation, where a case statement implements the decode block:

```
architecture Behavioral of bus_intfc is

    signal data_wr_d1            : std_logic_vector(7 downto 0);
    signal uart_tx_go_decode     : std_logic;
    signal fifo_read_decode      : std_logic;
    signal uart_tx_go_decode_d1  : std_logic;
    signal fifo_read_decode_d1   : std_logi

begin

    writes : process(clk)
    begin
        if rising_edge(clk) then
            data_wr_d1 <= data_wr;
            --
            uart_tx_go_decode <= '0';
            fifo_read_decode  <= '0';
            memory_wr         <= '0';

            if (wr_en = '1') then
                case (address) is
                    when X"001"    => uart_tx_go_decode <= '1';
                    when X"004"    => fifo_read_decode  <= '1';
                    when X"006"    => leds <= data_wr(3 downto 0);
                    when others    =>    --memory
                                    if (address(11) = '1') then memory_wr <= '1';
                                    else null;
                                    end if;
                end case;
            end if;
        end if;
    end process;
```

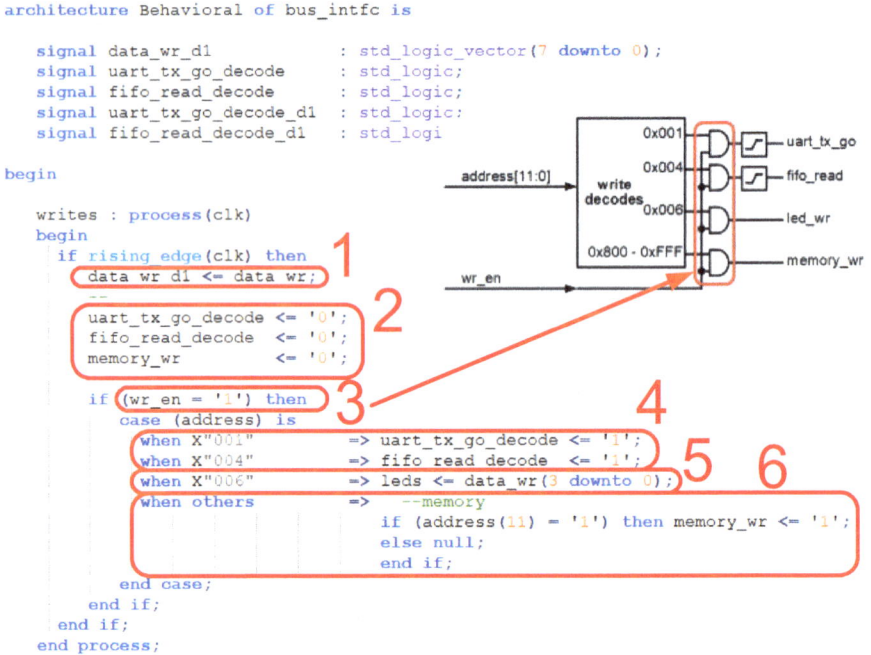

Fig. 6-14

1) Decoding the addresses takes one clock cycle (later in this process), and we delay the write data one clock here so that it lines up with the decoded write commands;

2) We activate these signals at specific times via the case statement below, and these "pre" zero assignments ensure that the signals go back inactive after the case assignments sets them active (we saw this operation previously);

3) The case statement is globally enabled with the write enable signal. This corresponds to the AND gates as shown;

4) These UART command signals will be active as long as "wr_en" is, and since we don't know how long the microprocessor will keep "wr_en" active (and since the UARTs require these command signals to be just one clock wide), as we'll see, we limit them to one clock farther down;

5) Data written to the external LEDs;

6) Here we get a little tricky. We want to write to the memory when the write-associated address is anywhere between 0x800 and 0xFFF. The "when others" case entry selects occurs any address other than the ones specifically decoded above, and this, of course, includes our memory range. From there, we need only then select just those addresses with the MSB active (i.e., the upper half of the full twelve-bit memory-mapped bus address range).

This is the process that limits the UART command signals to just one clock:

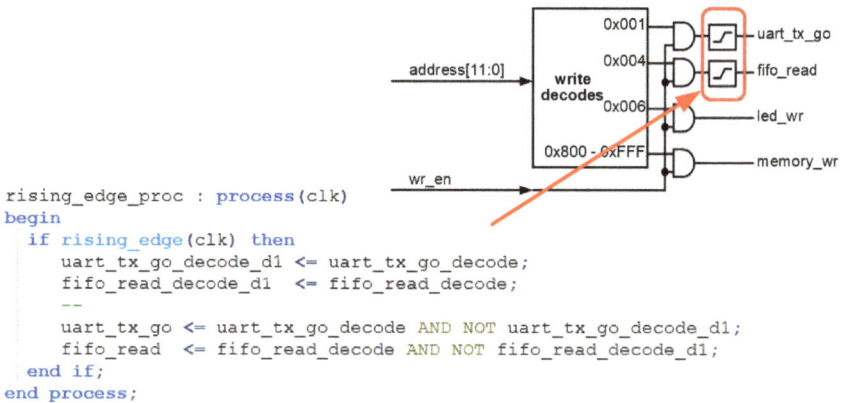

```
rising_edge_proc : process(clk)
begin
  if rising_edge(clk) then
     uart_tx_go_decode_d1 <= uart_tx_go_decode;
     fifo_read_decode_d1  <= fifo_read_decode;
     --
     uart_tx_go <= uart_tx_go_decode AND NOT uart_tx_go_decode_d1;
     fifo_read  <= fifo_read_decode AND NOT fifo_read_decode_d1;
  end if;
end process;
```

Fig. 6-15

... the rising-edge detection that we saw earlier.

And, finally, here's the reads, again using a case statement for the decodes:

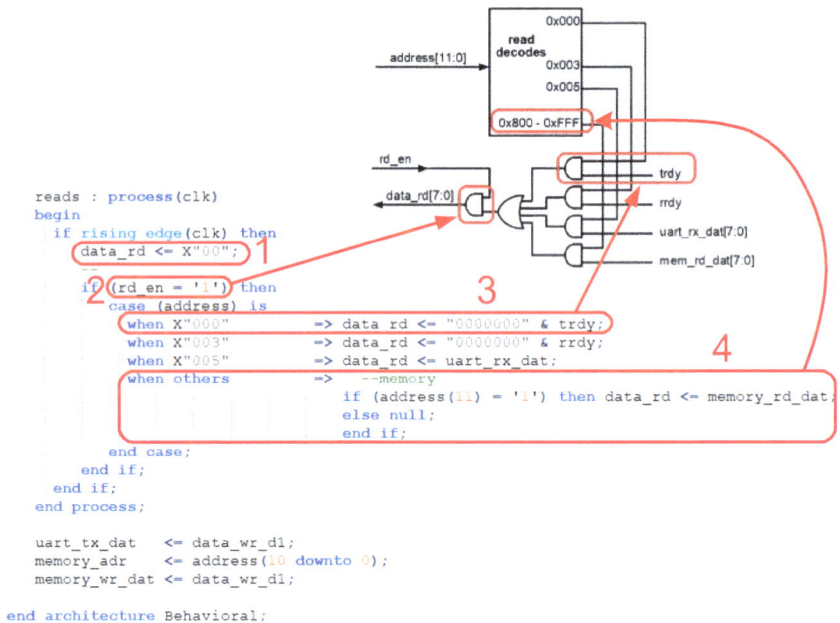

Fig. 6-16

1) The same "pre" clear as we saw with the writes,

2) but now the case statement is enabled by the read enable;

3) "trdy" (like "rrdy") is a one-bit read, and so it is concatenated (via the "&" symbol) with seven most-significant zeros;

4) Like the memory writes, here memory reads are implemented using the "when others" case selection.

Example Design Simulation Testbench

We can simulate this memory-mapped bus shell design with a testbench that simulates microprocessor writes and reads on the bus.

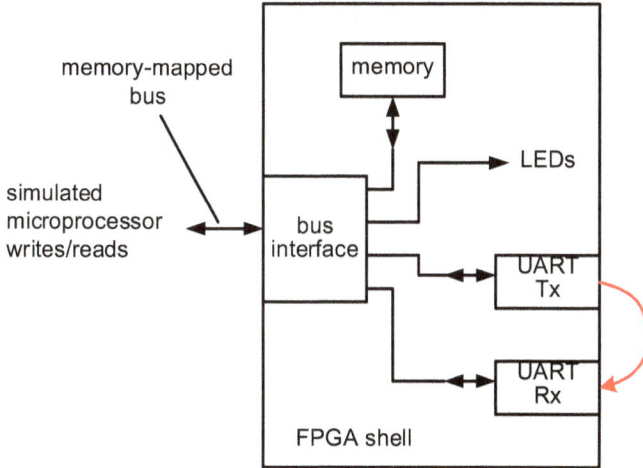

Fig. 6-17

Additionally, we can loop the UART Tx output back to the UART Rx input, saving the testbench the task of creating a simulated serial UART transmission.

The only component instantiated within the testbench is the top-level shell:

```
--   Push Button Testbench
-----------------------------------------------------------------

library IEEE;
use IEEE.STD_LOGIC_1164.all;
use IEEE.NUMERIC_STD.all;
use IEEE.STD_LOGIC_MISC.all;
use IEEE.STD_LOGIC_UNSIGNED.all;

entity shell_tb is
end entity shell_tb;

architecture Testbench of shell_tb is                          1
    -- reduced for simulation
    constant BAUD_9600 : std_logic_vector(15 downto 0) := X"0010";

    component shell is
    generic (BAUD_9600 : std_logic_vector(15 downto 0));       2
    port (
        clk          : in   std_logic;
        address      : in   std_logic_vector(11 downto 0);
        wr_en        : in   std_logic;
        data_wr      : in   std_logic_vector(7 downto 0);
        rd_en        : in   std_logic;
        data_rd      : out  std_logic_vector(7 downto 0);
        --
        uart_tx_sig  : out  std_logic;
        uart_rx_sig  : in   std_logic;
        leds         : out  std_logic_vector(3 downto 0)
        );
    end component;

    signal run_counter     : std_logic_vector(11 downto 0) := X"000";
    --UUT inputs (generated by the testbench)
    signal clk             : std_logic := '0';
    signal address         : std_logic_vector(11 downto 0) := X"000";
    signal wr_en           : std_logic;
    signal data_wr         : std_logic_vector(7 downto 0) := X"00";
    signal rd_en           : std_logic := '0';
    signal rrdy            : std_logic;
    signal uart_rx_sig     : std_logic;
    -- UUT outputs (observed by the testbench)
    signal data_rd         : std_logic_vector(7 downto 0);
    signal uart_tx_sig     : std_logic;
    signal leds            : std_logic_vector(3 downto 0) := X"0";

begin
```

Fig. 6-18

1) We set the BAUD rate extremely low for simulation;

2) This BAUD constant is then passed down via the shell's generic.

This shows the functional operation code of the testbench architecture section:

```
begin

    clk  <= '0' after 10 ns when clk = '1' else   -- 50MHz = 20ns
             '1' after 10 ns;

run_sequence : process
    variable i : integer := 0;
begin
    for i in 1 to 3 loop wait until rising_edge(clk); end loop;
    address <= X"006"; --LEDs
    data_wr <= X"05";
    wr_en   <= '1';
    wait until rising_edge(clk);
    wr_en   <= '0';
    --
    for i in 1 to 3 loop wait until rising_edge(clk); end loop;
    address <= X"800"; --memory
    data_wr <= X"AA";
    wr_en   <= '1';
    wait until rising_edge(clk);
    wr_en   <= '0';
    -- write to memory
    for i in 1 to 7 loop wait until rising_edge(clk); end loop;
    address <= X"921"; --memory
    data_wr <= X"BB";
    wr_en   <= '1';
    wait until rising_edge(clk);
    wr_en   <= '0';
    -- read from memory
    for i in 1 to 7 loop wait until rising_edge(clk); end loop;
    address <= X"800"; --memory
    wait until rising_edge(clk);
    rd_en   <= '1';
    for i in 1 to 2 loop wait until rising_edge(clk); end loop;
    rd_en   <= '0';
    for i in 1 to 2 loop wait until rising_edge(clk); end loop;
    address <= X"921"; --memory
    wait until rising_edge(clk);
    rd_en   <= '1';
    for i in 1 to 2 loop wait until rising_edge(clk); end loop;
    rd_en   <= '0';
    --
    for i in 1 to 8 loop wait until rising_edge(clk); end loop;
    address <= X"001"; --UART Tx
    data_wr <= X"09";
    wr_en   <= '1';
    for i in 1 to 4 loop wait until rising_edge(clk); end loop;
    wr_en   <= '0';
    --
    loop
        for i in 1 to 30 loop wait until rising_edge(clk); end loop;
        address <= X"003"; --rrdy
        rd_en   <= '1';
        for i in 1 to 2 loop wait until rising_edge(clk); end loop;
        if (data_rd(0) = '1') then
            wait until rising_edge(clk);
            address <= X"004"; --uart rx FIFO read
            wr_en   <= '1';
            for i in 1 to 4 loop wait until rising_edge(clk); end loop;
            wr_en   <= '0';
            wait until rising_edge(clk);
            address <= X"005"; --uart_rx_dat
            rd_en   <= '1';
            for i in 1 to 2 loop wait until rising_edge(clk); end loop;
            rd_en   <= '0';
        end if;
        wait until rising_edge(clk);
        rd_en   <= '0';
    end loop;
end process;
```

1

2

zoom

Fig. 6-19

1) This long process statement creates the entire sequence of stimulus signals for the simulation. With the rising-edge clocked process statements we've been using, the process only "wakes up" at each instant of the clock's occurrence. Then, all the assignments of the process statement essentially occur simultaneously. However, rising-edge process statements are the exception (although also the vast majority of instances) within the VHDL rules universe. The assignments of non-rising edge process statements like this one occur in sequence, each line in turn;

2) We'll zoom in on this area to explore further.

Rather than using a clocked counter and case statement to determine when to create the stimulus outputs at appropriated clock times, this process statement proceeds more or less in real time, meaning that we encounter each clock as it occurs in time:

```
run_sequence : process
  variable i : integer := 0;    5
begin
  for i in 1 to 3 loop wait until rising_edge(clk); end loop;
  address <= X"006"; --LEDs
  data_wr <= X"05";              1
  wr_en   <= '1';
  wait until rising_edge(clk);   2
  wr_en   <= '0';                3
  --
  for i in 1 to 3 loop wait until rising_edge(clk); end loop;   4
  address <= X"800"; --memory
  data_wr <= X"AA";
  wr_en   <= '1';
  wait until rising_edge(clk);
  wr_en   <= '0';
  -- write to memory
  for i in 1 to 7 loop wait until rising_edge(clk); end loop;
  address <= X"921"; --memory
  data_wr <= X"BB";
  wr_en   <= '1';
  wait until rising_edge(clk);
  wr_en   <= '0';
  -- read from memory
```

Fig. 6-20

1) For example, these three assignments are made essentially simultaneously (although technically in sequence),

2) and then the process pauses and waits for the next rising clock edge,

3) after which, this wr_en assignment is made,

4) and again we wait for a next clock edge. However, the "for" loop causes the testbench to wait through three clocks before proceeding.

5) We're introducing the concept of a variable, which is here used as the for-loop index counter. In VHDL, a variable inside a process statement is "local," meaning that it is only visible inside the process statement (unlike signals, which are visible—i.e., they exist—everywhere in the architecture and entity I/O). We could, for example, declare and use a variable with the same name in a different process statement, and they would be mutually exclusive—they would be two completely different variables. Note that, like other programming languages, we can reuse the "i" variable from one for-loop to the next. Although not done here (since we're using the "i" variable only in the for-loops), we assign variables via ":=" instead of the "<=" of a signal.

We'll zoom in on this area next:

```
begin

    clk  <= '0' after 10 ns when clk = '1' else   -- 50MHz = 20ns
             '1' after 10 ns;

    run_sequence : process
        variable i : integer := 0;
    begin
        for i in 1 to 3 loop wait until rising_edge(clk); end loop;
        address <= X"006"; --LEDs
        data_wr <= X"05";
        wr_en   <= '1';
        wait until rising_edge(clk);
        wr_en   <= '0';
        --
        for i in 1 to 3 loop wait until rising_edge(clk); end loop;
        address <= X"800"; --memory
        data_wr <= X"AA";
        wr_en   <= '1';
        wait until rising_edge(clk);
        wr_en   <= '0';
        -- write to memory
        for i in 1 to 7 loop wait until rising_edge(clk); end loop;
        address <= X"921"; --memory
        data_wr <= X"BB";
        wr_en   <= '1';
        wait until rising_edge(clk);
        wr_en   <= '0';
        -- read from memory
        for i in 1 to 7 loop wait until rising_edge(clk); end loop;
        address <= X"800"; --memory
        wait until rising_edge(clk);
        rd_en   <= '1';
        for i in 1 to 2 loop wait until rising_edge(clk); end loop;
        rd_en   <= '0';
        for i in 1 to 2 loop wait until rising_edge(clk); end loop;
        address <= X"921"; --memory
        wait until rising_edge(clk);
        rd_en   <= '1';
        for i in 1 to 2 loop wait until rising_edge(clk); end loop;
        rd_en   <= '0';
        --
        for i in 1 to 6 loop wait until rising_edge(clk); end loop;
        address <= X"001"; --UART Tx
        data_wr <= X"09";
        wr_en   <= '1';
        for i in 1 to 4 loop wait until rising_edge(clk); end loop;
        wr_en   <= '0';
        --
        loop
            for i in 1 to 30 loop wait until rising_edge(clk); end loop;
            address <= X"003"; --rrdy
            rd_en   <= '1';
            for i in 1 to 2 loop wait until rising_edge(clk); end loop;
            if (data_rd(0) = '1') then
                wait until rising_edge(clk);
                address <= X"004"; --uart rx FIFO read
                wr_en   <= '1';
                for i in 1 to 4 loop wait until rising_edge(clk); end loop;
                wr_en   <= '0';
                wait until rising_edge(clk);
                address <= X"005"; --uart_rx_dat
                rd_en   <= '1';
                for i in 1 to 2 loop wait until rising_edge(clk); end loop;
                rd_en   <= '0';
            end if;
            wait until rising_edge(clk);
            rd_en   <= '0';
        end loop;
    end process;
```

zoom

Fig. 6-21

Zoomed:

Fig. 6-22

1) Here, we loop with no conditions, meaning that we loop endlessly. As we'll soon see when we look at the simulated waveform, at this point in the simulation we're done testing the LEDs and memory, and this loop now provides continual monitoring of the Rx UART's rrdy status. We note that without this endless loop, the process statement would start over at the beginning (process statements are essentially themselves loops);

2) We wait 30 clocks between each rrdy status check, and then read the Rx UART's rrdy status;

3) We wait a couple of clocks to ensure that the rrdy bit has been placed on the memory-mapped read bus;

4) If the rrdy is active (meaning that the Rx UART's FIFO has at least one value), we'll proceed to get the received UART byte;

5) We read the FIFO, i.e., we execute an Rx UART read command;

6) And, finally, we read the value presented by the Rx UART (we observe the value that now appears in the simulation waveform).

The final piece of the testbench architecture is the shell module instantiation:

```vhdl
begin

    clk   <= '0' after 10 ns when clk = '1' else   -- 50MHz = 20ns
             '1' after 10 ns;

    run_sequence : process
        variable i : integer := 0;
    begin
        for i in 1 to 3 loop wait until rising_edge(clk); end loop;
        address <= X"006"; --LEDs
        data_wr <= X"05";
        wr_en   <= '1';
        wait until rising_edge(clk);
        wr_en   <= '0';
        --
        for i in 1 to 3 loop wait until rising_edge(clk); end loop;
        address <= X"800"; --memory
        data_wr <= X"AA";
        wr_en   <= '1';
        wait until rising_edge(clk);
        wr_en   <= '0';
        -- write to memory
        for i in 1 to 7 loop wait until rising_edge(clk); end loop;
        address <= X"921"; --memory
        data_wr <= X"BB";
        wr_en   <= '1';
        wait until rising_edge(clk);
        wr_en   <= '0';
        -- read from memory
        for i in 1 to 7 loop wait until rising_edge(clk); end loop;
        address <= X"800"; --memory
        wait until rising_edge(clk);
        rd_en   <= '1';
        for i in 1 to 2 loop wait until rising_edge(clk); end loop;
        rd_en   <= '0';
        for i in 1 to 2 loop wait until rising_edge(clk); end loop;
        address <= X"921"; --memory
        wait until rising_edge(clk);
        rd_en   <= '1';
        for i in 1 to 2 loop wait until rising_edge(clk); end loop;
        rd_en   <= '0';
        --
        for i in 1 to 5 loop wait until rising_edge(clk); end loop;
        address <= X"001"; --UART Tx
        data_wr <= X"09";
        wr_en   <= '1';
        for i in 1 to 4 loop wait until rising_edge(clk); end loop;
        wr_en   <= '0';
        --
        loop
            for i in 1 to 30 loop wait until rising_edge(clk); end loop;
            address <= X"003"; --rrdy
            rd_en   <= '1';
            for i in 1 to 2 loop wait until rising_edge(clk); end loop;
            if (data_rd(0) = '1') then
                wait until rising_edge(clk);
                address <= X"004"; --uart rx FIFO read
                wr_en   <= '1';
                for i in 1 to 4 loop wait until rising_edge(clk); end loop;
                wr_en   <= '0';
                wait until rising_edge(clk);
                address <= X"005"; --uart_rx_dat
                rd_en   <= '1';
                for i in 1 to 2 loop wait until rising_edge(clk); end loop;
                rd_en   <= '0';
            end if;
            wait until rising_edge(clk);
            rd_en   <= '0';
        end loop;
    end process;

    shell_inst : shell
    generic map (BAUD_9600 => BAUD_9600)
    port map
    (
        clk          => clk,
        address      => address,
        wr_en        => wr_en,
        data_wr      => data_wr,
        rd_en        => rd_en,
        data_rd      => data_rd,
        --
        uart_tx_sig  => uart_tx_sig,
        uart_rx_sig  => uart_tx_sig,
        leds         => leds
    );

end architecture Testbench;
```

Fig. 6-23

We won't zoom here, since the connections are simply one-for-one

Example Design Simulation Waveform

Here's the waveform from a run of the simulation, through LED and memory write/reads, and one transmission between the Tx and Rx UARTs:

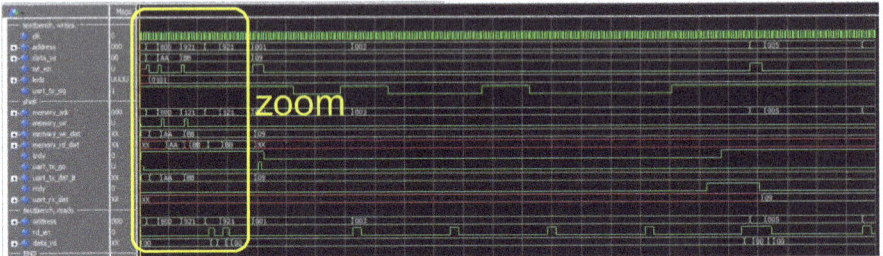

Fig. 6-24

We'll zoom in on the LED and Memory accesses first:

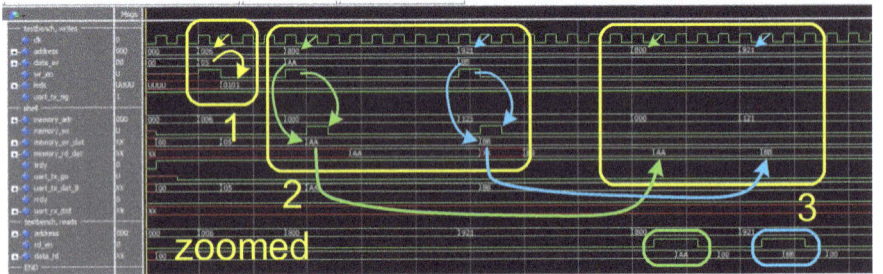

Fig. 6-25

1) We write to the LEDs, and see the proper bits driving them;

2) We write values to two different memory locations,

3) and read them back.

Lastly, we look at the Tx UART-to-Rx UART transmission:

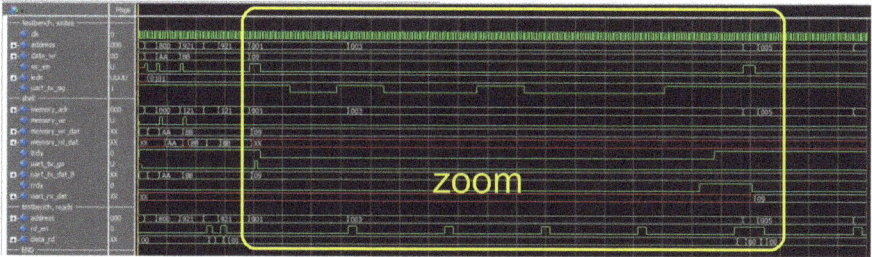

Fig. 6-26

Zoomed:

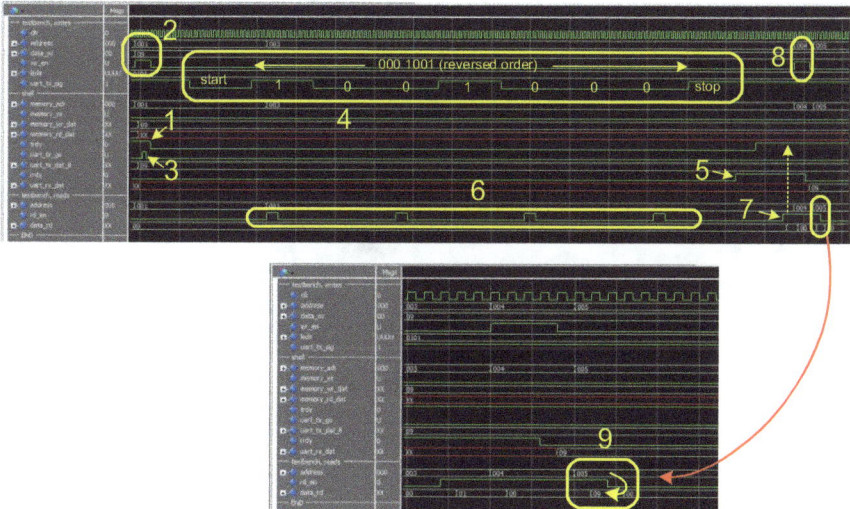

Fig. 6-27

1) The Tx UART's trdy status is good to go;

2) The simulated uProcessor writes a 0x09 to address 0x001 (UART Tx),

3) and then activates the uart_tx_go to initiate the transmission;

4) The transmission proceeds;

5) The Rx UART's rrdy status goes high, indicating that there's a received value ready in the FIFO;

6) The simulated uProcessor has been continually reading the Rx UART's FIFO status,

7) and at the next cycling Rx FIFO status read, the status is now active,

8) so the uProcessor issues the FIFO read command (address 0x004);

9) And we see that the received data is 0x09, the same as that which was transmitted.

Chapter 7

Odds and Ends

As I wrote in the introduction, the intent of this book is to get you off and running with VHDL. There's a lot still to be discovered, though. Fortunately, the internet (and specifically Wikipedia) has much to offer. Just knowing that something exists is often the key to a solution, and the following can perhaps be viewed as signposts to those solutions. Think of the following as a menu of offerings, where the entrees follow from other sources.

On the other hand, you may decide that at this point you'd like to tackle VHDL in a formal rigorous approach, in which case you'll perhaps want to spring for one (or more) of those thick, expensive texts.

abs
Returns the absolute value.

file
A VHDL type used for handling files, almost exclusively in simulation (versus synthesized logic)—allows stimulus vectors and simulation results to be accessed and stored in support files. Very powerful.

function
A function in VHDL is similar to those in other programming languages: a function can have one or more inputs, but just one virtual output—the value returned by the function. Functions are

called from an expression, and the returned value becomes part of that expression.

generate
Generally used to instantiate a replicated group of component modules, for example memory blocks. Arrays are typically used to map bit vector values to I/O ports.

package
Packages can be thought of as a type of include file. Generally, a package will contain declarations (e.g., arrays and records) and subprograms (generally functions) that are used in multiple modules, thus avoiding repeating them each place. Each module declares the package to be included, located above the beginning of the entity declaration (i.e., with the library declarations):

```
use work.pkg.all;
```
where in this case the name of the package is "pkg" (it could have been "frog"), and its file name would be "pkg.vhd".

parenthesis
We should note that many of the parenthesis used in the examples of the book are superfluous, included for clarity. Thus,

```
if (some_signal = '1') then
```
could be written as,

```
if some_signal = '1' then
```

PLLs
Stands for "Phase-Locked Loops". In an FPGA, a PLL can be thought of as a clock generator that is able to phase-lock to another clock (often an input to the FPGA). A PLL can generate clocks that are fractional multiples of the driving clock, e.g., four times, or one half, or even two-thirds (or often any integer fraction).

procedure
Where functions are similar to counterparts of the same name in other programming languages, procedures could be compared to subroutines. Unlike functions, procedures can have multiple

outputs, or even no outputs (or even no inputs). While called functions are placed in code where they are replaced by a single calculated value, procedures are typically called as a stand-alone expression to do potentially many different things, virtually anything in fact that the calling code could do—implement state machines, call other sub-modules, emulate a micro-processor.

record
Similar to an array, except instead of a table of similar-sized and typed values, a record can hold entries of different sizes and types. This is particularly useful when connecting a multitude of signals up and/or down through a multi-layered hierarchy, essentially bundles the group of signals together into one neat label.

rem
Stands for "remainder", and essentially returns the remainder after dividing one value by another. Thus, (X"7" rem X"3") would equal 1 (7 divided by 3 = 2, with 1 left over). The **mod** ("modulus") operator is similar to rem, but works with signed values as well.

resets
For the sake of clarity we haven't included resets in the examples of the book, but virtually all FPGA designs will include one or more. Further, most current FPGA resets will be clock-synchronous. A reset named "rst" might be used like this:

```
reset_example : process(clk)
begin
    if rising_edge(clk) then
        if (rst = '1') then
            some_signal <= '0';
        else
            [the usual logic operations]
        end if;
    end if;
end process;
```

RTL

Stands for "Register Transfer Logic", and is synonymous generically with HDL languages (e.g., VHDL).

shifting

We performed shifts by assigning offset bit vector values, but VHDL supports shift expressions:

sll: shift left, logical—fills LSB zeros;
srl: shift right, logical—fills MSB zeros;
sla: shift left, arithmetic—fills with current LSB;
sra: shift right, arithmetic—fills with current MSB.

subtype

Simply a subset of another type.

transceivers

Most high-end FPGA devices now include high-speed serial transceivers as build-in IP. Operating at multi-gigabit rates, they can transfer vast amounts of data between devices. Firewire, Ethernet, and PCIe are early examples.

We should also recognize that, while memory-mapped buses (with parallel, bit-vector data and address signals) are used internally in FPGAs (e.g., the ARM processor AXI bus), they are now rarely used externally, other than for the occasional out-board memory access.

x

Not to be confused with the "X" of hexadecimal types (e.g., X"5A"), "x" is a bit type that indicates an unknown value. Thus, "10xx" could indicate any value from 8 to 11 ("1000" to "1011").

z

A 'z' indicates a high-impedance value. Often used to define a tri-state output (or bi-directional signal) of an FPGA pin.

Index

www.ingramcontent.com/pod-product-compliance
Lightning Source LLC
Chambersburg PA
CBHW050652270326
41927CB00012B/2988